Lecture Notes in Artificial Intelligence   1623

Subseries of Lecture Notes in Computer Science
Edited by J. G. Carbonell and J. Siekmann

Lecture Notes in Computer Science

Edited by G. Goos, J. Hartmanis and J. van Leeuwen

Springer
Berlin
Heidelberg
New York
Barcelona
Hong Kong
London
Milan
Paris
Singapore
Tokyo

Thomas Reinartz

Focusing Solutions for Data Mining

Analytical Studies and Experimental Results
in Real-World Domains

 Springer

Series Editors

Jaime G. Carbonell, Carnegie Mellon University, Pittsburgh, PA, USA
Jörg Siekmann, University of Saarland, Saarbrücken, Germany

Author

Thomas Reinartz
DaimlerChrysler AG, Research and Technology
Wilhelm-Runge-Straße 11, D-89081 Ulm, Germany
E-mail: thomas.reinartz@daimlerchrysler.com

Cataloging-in-Publication data applied for

Die Deutsche Bibliothek - CIP-Einheitsaufnahme

Thomas Reinartz:
Focusing solutions for data mining : analytical studies and experimental results
in real-world domains / Thomas Reinartz. - Berlin ; Heidelberg ; New York ;
Barcelona ; Hong Kong ; London ; Milan ; Paris ; Singapore ; Tokyo : Springer, 1999
   (Lecture notes in computer science ; 1623 : Lecture notes in artificial intelligence)
   ISBN 3-540-66429-7

CR Subject Classification (1998): I.2, F.2, H.3, J.1, J.2

ISBN 3-540-66429-7 Springer-Verlag Berlin Heidelberg New York

© Springer-Verlag Berlin Heidelberg 1999
Printed in Germany

Typesetting: Camera-ready by author
SPIN 10705288   06/3142 – 5 4 3 2 1 0   Printed on acid-free paper

To My Family

Preface

This dissertation develops, analyzes, and evaluates focusing solutions for data mining. Data mining is a particular phase in knowledge discovery that applies learning techniques to identify hidden information from data, whereas knowledge discovery is a complex, iterative, and interactive process which covers all activities before and after data mining. Focusing is a specific task in the data preparation phase of knowledge discovery. The motivation of focusing is the existence of huge databases and the limitation of data mining algorithms to smaller data sets. The purpose of focusing is data reduction before data mining, either in the number of tuples, the number of attributes, or the number of values. Then, data mining applies techniques to the reduced data and is still able to achieve appropriate results.

In this dissertation, we first analyze the knowledge discovery process in order to understand relations between knowledge discovery tasks and focusing. We characterize the focusing context which consists of a data mining goal, data characteristics, and a data mining algorithm. We emphasize classification goals, top down induction of decision trees, and nearest neighbor classifiers. Thereafter, we define focusing tasks which include evaluation criteria for focusing success. At the end of the first block, we restrict our attention to focusing tasks for the reduction of the number of tuples.

We start the development of focusing solutions with an analysis of state-of-the-art approaches. We define a unifying framework that builds on three basic techniques: Sampling, clustering, and prototyping. We describe instantiations of this framework and examine their advantages and disadvantages. We follow up the unifying framework and establish an enhanced unified approach to focusing solutions which covers two preparation steps, sorting and stratification, and the application of sampling techniques. We reuse random sampling and systematic sampling from statistics and propose two more intelligent sampling techniques, leader sampling and similarity-driven sampling. We implement the unified approach as a generic sampling algorithm and integrate this algorithm into a commercial data mining system.

Thereafter, we analyze and evaluate specific focusing solutions in different domains. We exemplify an average case analysis to estimate expected average classification accuracies of nearest neighbor classifiers in combination with simple random sampling. We further conduct an experimental study and consolidate its results as focusing advice which provides heuristics for appropriate selections of best suited focusing solutions. At the end, we summarize the main contributions of this dissertation, describe more related work, raise issues for future work, and state some final remarks.

Contents

List of Figures

List of Tables

Chapter 1

Introduction

In this chapter, we introduce knowledge discovery and data mining. We motivate and characterize the primary goal of this dissertation, focusing solutions for data mining. Finally, we outline structure and contents of this dissertation, and briefly summarize its main contributions.

1.1 Knowledge Discovery in Databases and Data Mining

Today's information technology enables elicitation, storage, and maintenance of huge data volumes. The amount of data collections, databases, and data warehouses in all industries grows at phenomenal rates. From financial sectors to telecommunications, companies increasingly rely on analysis of data to compete. Although ad hoc mixtures of statistical techniques and file management tools once sufficed for digging through corporate data, the size of modern data warehouses, the mission-critical nature of data, and the speed with which companies need to make and update analyses call for a new approach (Brachman *et al.*, 1996). A new generation of techniques and tools is currently emerging to intelligently assist humans in analyzing huge data volumes and finding critical nuggets of useful knowledge. These techniques and tools are subject of the rapidly growing field of *Knowledge Discovery in Databases* (*KDD*) (Fayyad *et al.*, 1996).

Knowledge Discovery in Databases

By now, KDD is established as its own scientific area with its own conferences (Piatetsky-Shapiro, 1991; 1993; Fayyad & Uthurasamy, 1994; Fayyad *et al.*, 1995; Simoudis *et al.*, 1996; Heckerman *et al.*, 1997) and journal (Fayyad, 1997). KDD is an interdisciplinary field, and many related research areas contribute to KDD. Among them are *database technology*, *statistics*, and *machine learning* (Fayyad, 1996). Although the potential power of KDD increases with the high number of

T. Reinartz: Focusing Solutions for Data Mining, LNAI 1623, pp. 1-9, 1999
© Springer-Verlag Berlin Heidelberg 1999

contributions from many related fields, the wide range of research also results in
a labyrinth (Reinartz & Wirth, 1995). It is hard to find the most appropriate
approach among all fields, techniques, and tools to solve business challenges.
Likewise, it is also hard to define what KDD really is. A number of different
views and several attempts to define KDD exist. Within the last years, KDD
has been accepted as a complex process. The most prominent definition of KDD
is:

> "*Knowledge Discovery in Databases* is the non-trivial process of iden-
> tifying valid, novel, potentially useful, and ultimately understandable
> patterns in data." (Fayyad *et al.*, 1996)

This definition clearly states the complexity of KDD as a non-trivial process. It considers data as a set of facts and patterns as expressions in some
language describing subsets of data or models applicable to that subset. In this
definition, extracting patterns also designates fitting models to data, finding
structures from data, or in general any high-level description of a set of data.
The term process implies that KDD has many steps and multiple iterations.
The discovered knowledge should be valid on new data with some degree of
certainty and novel at least to the system, preferably to the user. The knowledge should also be potentially useful, i.e., lead to some benefits in the business.
Finally, discovered knowledge should be understandable, immediately or after
some post-processing. This definition is general, and its application in practice
requires exact definitions of all concepts.

Data Mining

One of the crucial steps in KDD is *data mining*. KDD and data mining are
often used as synonyms. It is common to use data mining in industrial contexts,
whereas KDD is more a scientific expression (John, 1997). In the scientific
sense, data mining corresponds only to a single phase in the KDD process. In
this dissertation, we adopt the distinction between KDD and data mining from
the scientific literature. In this sense, we use the following definition of data
mining:

> "*Data mining* is a step in the KDD process consisting of applying
> data analysis and discovery algorithms that, under acceptable com-
> putational efficiency limitations, produce a particular enumeration
> of patterns over the data." (Fayyad *et al.*, 1996)

Again, this definition is general and includes the application of any automatic
mechanism which is able to generate patterns from data. The only restriction
in this definition of data mining is the requirement of computational efficiency.
However, this definition does not specify what computational efficiency really
means. In practice, computational efficiency depends on the user and the requirements that result from the specification of business questions.

Research Challenges

Since KDD and data mining is a young discipline and data volumes across all industries grow by orders of magnitude from day to day, many research challenges still exist. These challenges include the following issues:

- *KDD Process*

  We still need a better understanding of the KDD process. We need to identify relevant phases in KDD, clarify relations among various steps during KDD processes, and define guidelines that help business analysts and data mining engineers to perform KDD projects.

- *Dynamic Aspects*

  The world of data is permanently changing. New data is entered into databases, behavior of customers varies over time, and discovered knowledge becomes obsolete (see Taylor & Nakhaeizadeh (1997), for example). KDD needs mechanisms to deal with all of these dynamic aspects. For example, monitoring processes to control whether discovered knowledge remains valid and to automatically (and preferably incrementally) adapt outdated data mining models yield significant benefits in comparison to static approaches.

- *Multi-Strategy Approaches*

  Single data mining algorithms are often not able to solve complex business questions. Moreover, KDD covers more than the data mining phase and consequently needs more than single techniques to perform entire KDD processes. Thus, we need the development of multi-strategy approaches that combine several techniques, either to utilize multiple techniques in order to answer single business questions or to solve more than one task. Given a specific context, multi-strategy approaches also require guidelines for the selection of most appropriate techniques and how to combine them most effectively.

- *Scaling up KDD and Data Mining*

  Gigabytes, terabytes, and petabytes of data are no longer fiction but reality. The size of today's databases requires efficient and effective data access. It also obligates methods to scale up data mining algorithms. This includes development of new data mining algorithms or adaptation of existing data mining algorithms, in order to cope with huge data volumes as well as mechanisms to reduce data before data mining in KDD.

This set of challenges in KDD only outlines a broad spectrum of relevant research questions. In summary, KDD is an emerging scientific area which is largely driven by industrial needs. The origin of KDD is the pressing demand for more intelligent facilities to analyze huge databases and to identify beneficial knowledge in business terms. From this point of view, the true challenge in

KDD is the transfer of scientific work into business life, filling the gap between research and practice. In this dissertation, we focus on a specific challenge that is both scientifically interesting and practically relevant.

1.2   Focusing for Data Mining

The high potential for identifying hidden knowledge in huge data volumes is also a pitfall. Huge data volumes potentially contain much information, and not all of this information is necessary to solve a given business question. In KDD, we are interested in revealing exactly this information which is relevant to solve the business question. Hence, KDD and data mining is an approach to *focusing* in a more general sense. We attempt to reduce the overall information hidden in data to this particular knowledge which is valuable in terms of business goals.

Moreover, the origin of most data mining algorithms is statistics or machine learning. Neither field really deals with huge data volumes yet. Usually, statistical and machine learning algorithms are not able to handle huge data volumes. They are often limited to fairly small sets of examples. The more advanced data mining algorithms are, usually the more restricted is their applicability.

In this dissertation, we consider this type of focusing on relevant information and analyze particular ways to reduce the overall available information effectively. We specifically focus our attention on the problem of limited applicability of data mining algorithms to huge data volumes. In principle, there are two main alternatives to tackle this challenge:

1.   Scale up Data Mining Algorithms

2.   Data Reduction

Scale up Data Mining Algorithms

The first alternative tries to solve the challenge of handling huge data volumes by developing new data mining algorithms or adapting existing ones. For instance, some approaches try to make use of emerging hardware technology like massively parallel machines which are becoming more popular. Hence, some developments endeavor to scale up data mining algorithms by changing existing sequential techniques into parallel versions. For example, PC4.5 (Li, 1997) is a parallel version of the decision tree algorithm C4.5 (Quinlan, 1993). Other attempts towards parallel data mining include work described by Al-Attar (1997), Darlington *et al.* (1997), Galal *et al.* (1997), Kufrin (1997), Mackin (1997), McLaren *et al.* (1997).

However, parallel machines are expensive and less widely available than single processor machines. Moreover, parallel data mining algorithms usually require more than just the availability of massively parallel machines. For example, they need some extra software to distribute components of parallel algorithms among processors of parallel machines. This strategy also does not apply to all existing data mining algorithms. Some of them are sequential in nature, and can not make use of parallel hardware. Finally, it is usually an enterprise to re-implement existing software, and it is often not worth spending this effort.

Data Reduction

The second alternative attempts to reduce data before applying data mining algorithms. Since huge data volumes are likely to contain more information than necessary to successfully apply data mining algorithms, it is often possible to reveal comparable results by applications of data mining algorithms to subsets of the entire data. Beyond the possibility to enable applications of data mining algorithms which are normally limited to smaller data sets, this strategy leads to the following additional advantages. Data mining on subsets of data speeds up hypotheses formulation and testing. We generate hypotheses on subsets of data and only use the entire data set for hypothesis testing. If data mining on subsets of data already yields sufficiently good results, we do not need to apply data mining algorithms to the entire data at all.

Data reduction covers a wide area of research. In general, this research tries to enable applications of data mining algorithms which are not applicable to the entire data set, or it attempts to make data mining more feasible, e.g., in terms of running time. In some cases, data mining results on subsets of data even outperform data mining on the entire data set (e.g., Caruana & Freitag, 1994). It is also possible to use data reduction and resulting subsets of data to calibrate parameters to optimal values, and then to apply the data mining algorithm with this optimal parameter setting to the entire data. If single subsets of data are usable for more than single data mining algorithms, we also compensate higher costs of data reduction and can apply more expensive data reduction techniques. Finally, we are also able to apply the same data mining algorithms to several reduced data sets and to combine the intermediate results to a final data mining output (e.g., Provost & Kolluri, 1997).

Focusing

In this dissertation, we contribute to data reduction rather than scaling up data mining algorithms. Early papers in KDD refer to data reduction as *focusing*:

> "The focusing component of a [knowledge] discovery system determines what data should be retrieved from the DBMS. This requires specifying which tables need to be accessed, which fields should be returned, and which or how many records should be retrieved."
> (Matheus *et al.*, 1993)

Subsequently, we consider focusing in this sense of data reduction. We attempt to contribute to the focusing component of a knowledge discovery system and to analyze potentials of focusing in KDD.

General Goals of Dissertation

The general goals of this dissertation include the following topics:

- *KDD Process*

  Understand the KDD process and identify those aspects that mainly influence the definition of focusing tasks and the development or selection of appropriate focusing solutions.

- *Focusing Tasks*

  Define focusing tasks in detail and select specific focusing tasks for further examination.

- *Focusing Solutions*

  Develop, analyze, and evaluate focusing solutions for data mining.

At this point, these goals remain generic. Consequently, one of the first issues in this dissertation comprises the identification of more specific goals.

Note, although it is also a general issue in KDD projects to decide whether focusing is necessary at all, we do not consider this question. Instead, we assume that focusing is either necessary or that focusing is part of the KDD process for any other reason.

1.3 Overview

This dissertation contains three blocks. Figure 1.1 presents an overview of the most important chapters and sections in each of these blocks. The first block refines the broad spectrum of research challenges in KDD and defines a specific goal for this dissertation in this context. The second block develops, analyzes, and evaluates solution approaches to this goal. The third block concludes with a summary of main contributions, a discussion of additional related work, and an outline of remaining challenges for future work.

- In chapter 2, we describe KDD in more detail. We present a brief description of a process model for KDD projects. This process model clarifies the complexity of KDD and depicts the role of focusing in KDD projects. We refine the overall focusing context to those aspects that are particularly relevant for subsequent considerations. The focusing context includes data mining goals, data characteristics, and data mining algorithms. We illustrate different data mining goals, define concepts of data characteristics, and outline some specific data mining algorithms. At the end of the second chapter, we restrict the focusing context to *classification* goals and select two data mining algorithms, *top down induction of decision trees* and *nearest neighbor classifiers*, for further examination.

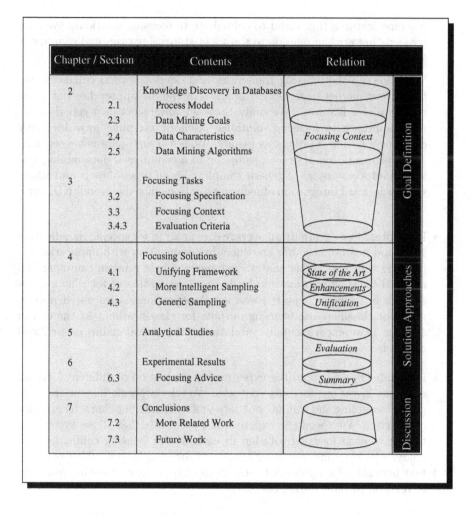

Chapter / Section		Contents	Relation	
2		Knowledge Discovery in Databases		Goal Definition
	2.1	Process Model		
	2.3	Data Mining Goals		
	2.4	Data Characteristics	*Focusing Context*	
	2.5	Data Mining Algorithms		
3		Focusing Tasks		
	3.2	Focusing Specification		
	3.3	Focusing Context		
	3.4.3	Evaluation Criteria		
4		Focusing Solutions		Solution Approaches
	4.1	Unifying Framework	*State of the Art*	
	4.2	More Intelligent Sampling	*Enhancements*	
	4.3	Generic Sampling	*Unification*	
5		Analytical Studies		
			Evaluation	
6		Experimental Results		
	6.3	Focusing Advice	*Summary*	
7		Conclusions		Discussion
	7.2	More Related Work		
	7.3	Future Work		

Figure 1.1: Structure and Contents of Dissertation

- In chapter 3, we specify a framework for the definition of focusing tasks. The definition of focusing tasks contains focusing specification and focusing context. Each focusing task needs different focusing solutions, and the context determines appropriate evaluation criteria for focusing success. At the end of the third chapter, we define specific evaluation criteria and select a subset of all possible focusing tasks. We focus our attention to focusing tuples on subsets of tuples rather than focusing attributes or values.

- In chapter 4, we start the development of focusing solutions for the selected focusing tasks with an analysis of state of the art approaches. We present a unifying framework that covers existing efforts to solve the selected focusing tasks. Thereby, we emphasize approaches in statistics and

machine learning that claim to contribute to focusing solutions. We analyze existing work in relation to the selected focusing tasks and summarize
advantages and disadvantages of existing methods.

We follow up the unifying framework and develop several enhancements
that overcome negative aspects of state of the art approaches but reuse
their positive features. We unify these enhancements in a generic sampling approach to focusing solutions. Generic sampling provides many
different focusing solutions as instantiations through parameter settings.
An implementation of generic sampling in a commercial data mining system enables easy usage of generic sampling for all types of users and allows
straightforward integration of focusing solutions into the overall KDD process.

- In chapter 5, we exemplify an average case analysis for specific classification
  goals and nearest neighbor classifiers in combination with applications of
  focusing solutions. This analysis compares simple random sampling and
  an ideal focusing solution and leads to an estimation of focusing success
  of both in terms of expected average classification accuracies, if nearest
  neighbor classifiers use focusing outputs for classification. At the end of
  chapter 5, we experimentally validate the theoretical claims on artificial
  data sets.

- In chapter 6, we extend the experimental evaluation of different focusing
  solutions in various focusing contexts. We select specific instantiations of
  generic sampling and a number of data sets with different characteristics for
  comparison. We specify an experimental procedure that allows systematic
  testing of each focusing solution in each of the focusing contexts. The
  analysis of experimental results leads to focusing advice which provides
  first heuristics for appropriate selections of best suited focusing solutions
  in relation to the focusing context.

- In chapter 7, we conclude with a summary of main contributions, a short
  discussion of more related work, and a description of issues for future work.

Reading Advice

In general, we assume that the reader of this dissertation is familiar with
principles and concepts in statistics, machine learning, and KDD. We shortly
recapitulate some standard definitions and approaches, and we introduce their
notations, if we either specifically use them, or if a basic understanding of these
concepts is necessary to clarify relations between existing work and our own
developments. For more detailed definitions and descriptions, we refer to introductory text books (e.g., Hartung & Elpelt, 1986; Hartung et al., 1987; Mason
et al., 1989; Langley, 1996; Mitchell, 1997; Adriaans & Zantinge, 1996; Fayyad
et al., 1996; Weiss & Indurkhya, 1997).

For our own work, we ensure that the reader is able to re-implement all concepts and algorithms. If too many details hinder readability, we only textually describe important aspects and precisely state formal contents in the appendix.

Chapter 2

Knowledge Discovery in Databases

In this chapter, we describe KDD as a complex, iterative, and interactive process. The KDD process specifies the overall context of focusing tasks and their solutions for data mining. Specific focusing contexts include data mining goal, data characteristics, and data mining algorithm. The primary goal of this chapter is clarification and refinement of the focusing context.

2.1 Knowledge Discovery Process

In the first chapter, we introduced KDD as a complex process and identified data mining as a single phase in KDD. However, the definition of KDD (see page 2) does not mention the important role of humans in the loop of KDD (Brachman & Anand, 1994; 1996). This definition also lacks precise specifications of KDD phases and tasks in order to guide business analysts and data mining engineers in performing KDD projects in practice.

In this section, we briefly describe four classes of humans in the loop of KDD and outline a process model for KDD. This process model defines the prerequisites to understand KDD as a complex, iterative, and interactive process and is a first step towards guidelines for KDD projects in practice. A more detailed understanding of KDD helps to identify crucial factors that influence the definition of focusing tasks and the development or selection of appropriate focusing solutions.

2.1.1 Humans in the Loop

Current views on KDD often neglect the important role of humans in the loop. KDD projects are likely to fail without support and intermediate decisions of

T. Reinartz: Focusing Solutions for Data Mining, LNAI 1623, pp. 11-44, 1999
© Springer-Verlag Berlin Heidelberg 1999

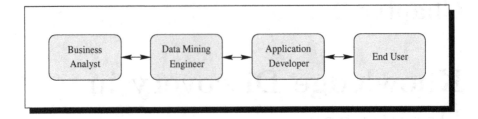

Figure 2.1: Humans in the Loop of KDD

humans. At different levels of granularity and at different levels of skills, human involvement is essential for the success of KDD projects and necessary for the effective usage of data mining algorithms in order to gain valuable information from raw data. In the following, we outline a human-centered view on KDD and discuss different roles of humans in the loop of KDD. Figure 2.1 presents an overview of four different classes of humans. KDD projects require participation of the following humans:

- *Business Analyst*

  At the beginning of KDD projects, the business analyst initiates the project by contacting key people in the business and summarizing the initial situation. The business analyst investigates the existing business, identifies business questions, considers current business solutions, and inspects potentials for KDD in terms of the application domain. His engagement covers the definition of more precise goals and their translation into KDD terminology. Business analysts are not necessarily sophisticated experts in KDD but they possess the necessary business background.

- *Data Mining Engineer*

  The data mining engineer starts with analyses of concrete data mining goals. He is able to access data and to prepare the original data for applications of data mining algorithms. He selects and applies appropriate techniques and generates data mining results. After several iterations with different techniques and different parameter settings, the data mining engineer summarizes the data mining results or creates a generic KDD procedure that solves the original business goal. Data mining engineers are familiar with data mining algorithms and know how to apply these techniques in order to solve data mining goals.

- *Application Developer*

  If the overall objective of the KDD project is not only the development of single data mining results but the integration of data mining solutions into the business, the application developer (team) realizes appropriate systems and integrates them into the business environment. Application developers are able to analyze existing technical and organizational conditions in order to develop final systems that implement solution approaches and fit into the existing environment.

- *End User*

  The resulting system abstracts from KDD specific know-how and details. It is easy for end users in the business to use the final system for their day-to-day analysis. End users are not familiar with KDD and data mining technologies but they are able to understand the business questions and to deploy data mining results into practice.

The first two classes of humans in the loop are fundamental for success of KDD projects, whereas the involvement of application developers and end users depends on the purpose of the project. If single data mining results are sufficient to meet the project objectives, there is no need for the participation of the third and fourth class of humans.

Humans involved in KDD projects do not really need to be different. The smaller the project is, the more likely a single person plays more than a single role in the project. Moreover, the efforts of humans in KDD usually overlap and sometimes require joint activities to achieve the overall goal of the project. If the KDD project is large, it is also possible to have several persons who take part in the project with similar roles.

In our context, we notice that focusing is an issue of the data mining engineer since focusing requires substantial skills in KDD technology and sufficient expertise in data mining algorithms. The business analyst only works on the prerequisites that support selection and application of appropriate data mining algorithms as well as identification of all necessary data preparation tasks.

2.1.2 KDD Project Phases

The focus of KDD projects is usually data mining and how to achieve data mining goals. However, the complete KDD process is not simply an application of highly sophisticated data mining algorithms and an immediate deployment of data mining results into the business. Instead, KDD projects perform an iterative process of various phases and tasks. Unfortunately, current research in KDD mainly emphasizes technical details of data mining algorithms or very specific applications. The relations between techniques from different fields and how they fit into the overall KDD process often remain unclear. For business analysts and data mining engineers, it is often hard to find the best suited techniques to solve the business question at hand. A coherent framework for KDD and a common methodology that helps practitioners to perform their KDD projects do not exist (Reinartz & Wirth, 1995; Wirth & Reinartz, 1995).

In this section, we propose a process model for KDD that guides business analysts and data mining engineers in applications of KDD technologies

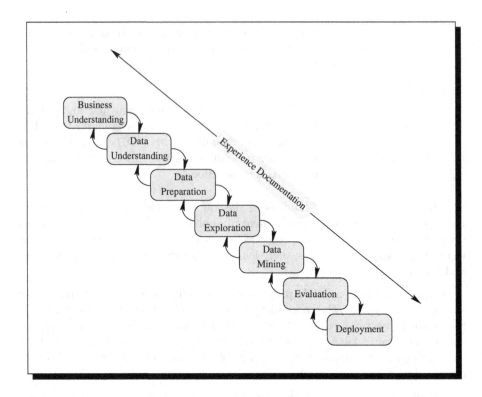

Figure 2.2: KDD Phases

(Nakhaeizadeh *et al.*, 1997). This process model follows up existing attempts towards a KDD methodology (Agrawal *et al.*, 1993; Carbone & Kerschberg, 1993; Brachman & Anand, 1994).

The process model for KDD is an overview of the life cycle of KDD projects at a generic level. It contains the corresponding phases of a project, their respective generic tasks, and relationships between these tasks. Here, we only outline the phase level of the process model. Generic tasks and their relationships are beyond the scope of this dissertation (Chapman *et al.*, 1998). For our purpose, a basic understanding of the KDD process and its phases is sufficient to analyze the focusing context and relations between other issues in KDD projects and focusing.

The life cycle of KDD projects consists of seven phases. Figure 2.2 shows all phases in an idealized sequential order from top left to bottom right. However, the sequence of phases is not strict. KDD projects always involve moving back and forth between different phases, and it depends on the outcome of each phase which phase, or which particular task of a phase, follows next. At generic levels, KDD projects include the following phases:

- *Business Understanding*

  The business understanding phase aims at the precise definition of project objectives, business questions, and their translation into data mining goals. An elicitation of requirements on the data mining results, the initial identification of domain factors, and the assessment of the feasibility of the project are part of this phase as well. The final result of business understanding are recommendations for further proceeding and the preparation of subsequent phases.

- *Data Understanding*

  The data understanding phase builds on business understanding. Starting with the data mining goals, the main objective of data understanding is the analysis and documentation of available data and knowledge sources in the business as well as the initial computation of basic data characteristics.

- *Data Preparation*

  The data preparation phase provides the data set for subsequent phases. Data preparation contains all tasks before the application of exploration and data mining algorithms. This includes initial data selection and focusing as well as transformation and cleaning of data. If necessary, this phase also covers the integration of different information sources. The data preparation phase is typically iterative and usually the most time-consuming part of KDD projects.

- *Data Exploration*

  In the data exploration phase, the main objective is to get more familiar with the data, to discover first insights into the data, or to identify interesting data subsets which lead to hypotheses for hidden information. This phase includes the evaluation of initial hypotheses or leads to an application of data mining algorithms in order to verify these hypotheses. The most useful tools exploited in this phase are descriptive statistics and visualization.

- *Data Mining* [1]

  The data mining phase covers selection and application of modeling and discovery techniques as well as calibration of their parameters to optimal values. Typically, several techniques exist for the same data mining goal, and the data mining phase includes applications of different techniques as well as local evaluation and comparison of results.

---

[1] At this point, we remind that KDD refers to the entire process from raw data to valuable knowledge whereas data mining only depicts a particular phase in KDD that covers the application of modeling and discovery algorithms. In industry, both terms are often used as synonyms for the entire process.

- *Evaluation*

  In the evaluation phase, KDD projects conduct interpretation and evaluation of data mining results in business terms. This phase identifies shortcomings and initiates new experiments if necessary. Beyond evaluation of data mining results, this phase covers the assessment of the whole process as well. For example, checking remaining resources and allocating new resources to exploit further potential for KDD are part of the evaluation phase. At the end, recommendations for alternative next steps guide further processing.

- *Deployment*

  Finally, the deployment phase transfers data mining results into the business, if the project results meet the success criteria. Depending on the requirements, the deployment phase is as simple as generating a report or as complex as realizing and integrating a data mining analysis environment. Frequently, deployment consists of the integration of predictive models into an existing environment. Maintenance and monitoring data mining results is also part of deployment.

- *Experience Documentation*

  Experience documentation during all phases of KDD projects ensures that all experiences of the project and of the deployment of data mining results are reported. This phase is strictly speaking not part of the solution for the particular business question. However, documentation of experiences is important since collected sets of experiences over time make the performance of subsequent projects more efficient and more effective.

Focusing is a task in data preparation. For this reason, we now consider data preparation in more detail and relate focusing to other tasks in data preparation. On the other hand, focusing partly depends on phases before data preparation and phases after data preparation. Consequently, we also consider aspects of business understanding, data understanding, and data mining. From a focusing perspective, these phases contain the most important tasks which influence the definition of focusing tasks and the development or selection of appropriate focusing solutions.

2.2   Data Preparation

Data preparation is usually the most time-consuming and challenging phase in KDD (Famili *et al.*, 1997). The estimated effort of data preparation in KDD projects is about 50% of the overall process (Simoudis, 1997). In this section, we outline data preparation tasks and consider the particular close relation between data selection and focusing.

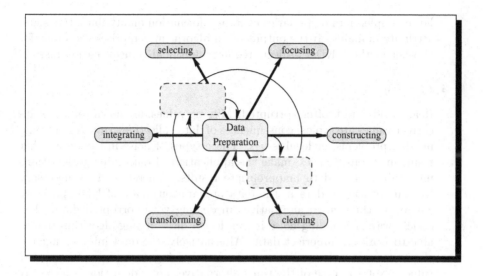

Figure 2.3: Data Preparation Tasks

2.2.1 From Business Data to Data Mining Input

If we assume raw databases that contain the relevant information at the beginning of KDD projects, the overall goal of data preparation is to construct a
data set that is suitable for a data mining algorithm as input. The data preparation phase covers all issues of manipulating data before the data exploration
and data mining phases. Most algorithms work on flat files, they all require
different formats, and they all have different characteristics. For example, some
data mining algorithms are only able to handle qualitative attributes whereas
others only work on quantitative attributes. Data preparation often happens
more than once. For instance, if the project plans to apply more than a single
data mining algorithm, each algorithm usually requires its own data set.

Figure 2.3 depicts all data preparation tasks in KDD. Beyond data selection
and focusing which we discuss later, data preparation comprises the following
tasks:

- *Transforming*

  In many KDD projects, databases contain the data for analysis. However, most existing data mining algorithms work on flat files. Hence, it
  is necessary to transform data from databases into regular ASCII files.
  Moreover, different data mining algorithms require different types of files.
  For example, some techniques expect blanks as separators between attribute values whereas others only work on comma-separated entries. For
  algorithms that explicitly need an extra file with information on attribute
  domains, the data mining engineer has to prepare this information as well.

For example, data dictionaries contain information on attribute types and attribute domains. If the output of an algorithm comprises the input for another method, KDD projects require additional transformation tasks.

- *Cleaning*

  Real-world data is often corrupted. Missing values, unknown values, or incorrect values are typical consequences of data collection errors, data transmission problems, or legal issues. These types of noise in the data either result in inconsistencies, make the application of data mining algorithms impossible, or lead to inappropriate results. Therefore, it is necessary to deal with noisy data in the data preparation phase of KDD projects. There are three main alternatives in dealing with corrupted data. The easiest way of handling noise is to select a data mining algorithm that is able to work on imperfect data. Alternatively, the data mining engineer eliminates noisy data or tries to correct missing, unknown, or incorrect values. Note, in case of the third alternative, noise detection is also part of cleaning in data preparation. Example approaches to cover the problem of missing values are described by Dempster *et al.* (1977), Little & Rubin (1987), Lakshminarayan *et al.* (1996).

- *Constructing*

  If the present data does not contain the necessary information to solve the given business question, additional data collection or data construction is necessary. For example, the data mining engineer buys additional data from professional providers or constructs new data by combining existing attributes or using constructive induction (e.g., Weiss & Kulikowski, 1991). Simple statistics often serve the same purpose for computing additional information beyond the initial raw data.

- *Integrating*

  Data integration is a specific data preparation task in KDD projects that aim to utilize information from different data or knowledge sources. In these cases, the data mining engineer integrates all information into a single data source which consolidates the input for subsequent phases in KDD. Data integration covers organizational issues as well as technical issues. For example, data mining engineers often apply table joins to construct a compound table which contains all information from several tables to solve the business question.

Since focusing is a small process in itself, and focusing solutions are also algorithms, these data preparation tasks are often necessary before applications of focusing solutions as well. Now, we consider data selection and focusing as the remaining tasks in data preparation.

2.2.2 Data Selection and Focusing

Databases usually contain more information than necessary to achieve specific business goals. Data selection and focusing both refer to data preparation tasks which allow for data reduction before the application of data mining algorithms. We distinguish between data selection and focusing in the following way.

Data selection refers to the selection of *target data*. For example, marketing databases contain data describing customer purchases, demographics, and lifestyle preferences. In order to identify which items and quantities to purchase for a particular store, as well as to answer the question how to organize items on the store's shelves, a marketing executive possibly only needs to combine customer purchase data with demographic data. Hence, he selects the respective data from the database which contains information on customer purchases and demographic data but neglects data on lifestyle preferences. On the other hand, focusing corresponds to automatic data reduction which does not necessarily rely on background information. For example, a marketing executive generates a simple random sample of a reasonable size, if the data on customer purchases and demographics still includes more data than the following algorithms are able to handle.

We conclude that data selection and focusing describe similar tasks, and that there is a close relation between them. In our understanding, there is a strong connection between data access and selection whereas focusing is more independent from data access. Data access and selection usually happen before focusing. Whereas data access concerns technical issues of querying databases, data selection often results in a first data reduction. For example, database queries to access data often contain WHERE clauses which restrict the entire database table to a specific subset. However, the selected data potentially still contains more information than necessary to achieve given data mining goals. Then, we need further data reduction, and focusing is a data preparation task following data selection.

Figure 2.4 shows part of the KDD process and illustrates the relation between data selection and focusing. We start KDD with a raw database. Then, we use data selection to specify the target data. Afterwards, we apply focusing solutions to reduce the target data again. The focusing output is then the input to the data mining phase. Since focusing is the central goal of this dissertation, we focus our attention on the limited process between selected data and data mining.[2] We concentrate our efforts on automatically focusing data, neither on data access or database interfaces nor on manually selecting relevant subsets of the entire database.

Forward Selection and Backward Elimination

We are generally able to perform focusing in two different directions. We

---

[2]Note, focusing solutions are also useful for different phases such as data exploration.

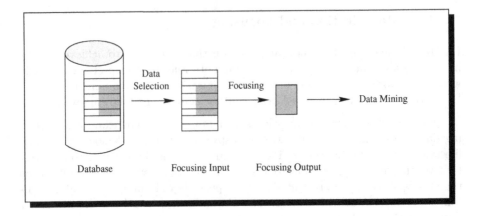

Figure 2.4: Data Selection and Focusing

either start with an empty focusing output and iteratively add elements of the original data set, or we begin focusing with an initial focusing output that contains all elements of the original data and iteratively remove elements. The first strategy is *forward selection*, whereas the second alternative is *backward elimination* (Caruana & Freitag, 1994; Wilson & Martinez, 1997a). We propose to follow the forward selection strategy.

Although focusing is a data preparation task and we focus our attention to the KDD process between data selection and data mining, focusing also depends on phases before data preparation. Therefore, we first step back in the process and consider crucial activities in business understanding and main characteristics in data understanding. Thereafter, we also step forward in the process and describe two specific data mining algorithms which we apply later.

2.3   Data Mining Goals

At the beginning of KDD projects, it is crucial for the final success of KDD to ensure precise definitions of data mining goals. Within the business understanding phase, the business analyst considers the application domain, identifies some pressing business questions, and translates the business questions into data mining goals. Usually, KDD projects involve sequences of different goals which together solve the original business questions. The data mining goal, or possibly a set of data mining goals, forms the core of KDD projects. From focusing perspectives, it is important to know which particular data mining goal is considered in the KDD process, since focusing tasks and the development or selection of appropriate focusing solutions highly depend on the data mining goal. In this section, we outline different data mining goals and describe classification goals more precisely.

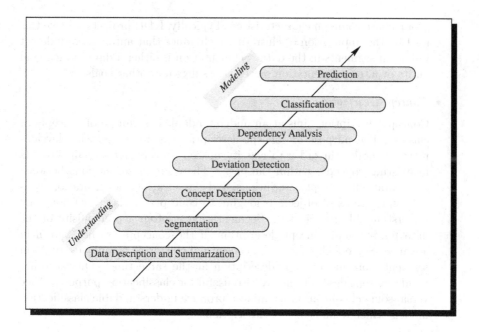

Figure 2.5: Data Mining Goals

2.3.1 From Understanding to Predictive Modeling

Data mining goals vary between two extremes. At the more vague end of data mining goals, the main purpose of KDD projects is *understanding* the nature of data and getting first insights into potential nuggets of knowledge. At the more concrete end of data mining goals, *predictive modeling* aims at the extraction of patterns that lead to some executable procedure and valuable returns in business terms. Figure 2.5 outlines the following most important data mining goals from understanding to predictive modeling (Nakhaeizadeh *et al.*, 1997):

- *Data Description and Summarization*

  Data description and summarization aims at concise descriptions of data characteristics, typically in elementary and aggregated form. These characteristics convey an overview of the structure of the data. Sometimes, data description and summarization is the only objective of KDD. However, in almost all projects data description and summarization is a subgoal, typically in early stages. Initial exploratory data analysis helps to understand the nature of the data and to find potential hypotheses for hidden information.

- *Segmentation*

  The data mining goal segmentation aims at the separation of data into interesting and meaningful subgroups or classes. All members of a subgroup share common characteristics. Typically, KDD projects achieve this goal by the application of clustering techniques that automatically detect unknown segments in the data. Segmentation is either a data mining goal on its own, or segmentation serves as a subgoal of other goals.

- *Concept Description*

  Concept description aims at an understandable description of concepts or classes. The purpose of concept descriptions is not necessarily development of models in order to classify unseen objects but to gain insights in existing concepts within the data. There are close relations between segmentation, concept descriptions, and classification. Segmentation leads to enumerations of objects belonging to concepts or classes without any understandable description. Segmentation is often a prerequisite of the data mining goal concept description, if the concepts are not known in advance. Some techniques, e.g., conceptual clustering techniques, perform segmentation and concept description at the same time. On the other hand, concept descriptions are also useful for classification purposes. Vice versa, some classification techniques produce understandable classification models which define concept descriptions.

- *Deviation Detection*

  An object (or set of objects) is a deviation, if it differs from norms or from expected values. As a data mining goal, deviation detection leads to the following implications. First, deviations indicate problems with data quality. In this case, we detect deviations and clean the resulting problems in data quality if necessary. Second, deviations are indicators for new phenomena unknown to the business. Then, we analyze deviations in more detail and identify additional data mining goals as a consequence.

- *Dependency Analysis*

  Dependency analysis consists of finding a model which describes significant dependencies between objects or events. For example, dependencies predict values of entities given information on other entities. *Associations* are a specific type of dependencies. They describe affinities of entities (i.e., objects or events which frequently occur together). Dependency analysis is closely related to classification and prediction. For both data mining goals, dependencies implicitly form predictive models. There is also a connection to concept descriptions which often highlight dependencies.

  Sequential patterns are similar to dependencies and associations. Whereas dependencies describe static events, sequential patterns cover sequences of events over time.

- *Classification*

  Classification assumes the existence of objects which belong to different classes. The goal in classification is to build classification models which assign the correct class to previously unseen and unlabeled objects. Classes are either given in advance, for instance by an end user, or derived from segmentation subgoals. Classification is one of the most important data mining goals that occurs in a wide range of various applications.

- *Prediction*

  The data mining goal prediction is similar to classification. The only difference is that the class in prediction is not symbolic but refers to numeric values. This means that the aim of prediction is to find the numeric value of an attribute for unseen objects. If prediction deals with time series data, it is also called *forecasting*.

Note, data mining goals do not exclude each other, and a single goal is often the prerequisite for another data mining goal. Some data mining goals also overlap with other goals. Moreover, a solution of the initial data mining goal often helps to identify the next data mining goal and its solution.

2.3.2 Classification

In the following, we concentrate on the data mining goal classification. We choose classification as the primary data mining goal since most applications correspond to classification tasks. For example, many business questions at Daimler-Benz including the detection of early indicator cars (Wirth & Reinartz, 1996) and credit scoring in automotive industries (Kauderer & Nakhaeizadeh, 1997) contain aspects of classification goals. Other examples include applications in diagnosis (see Goos (1995), for example) or satellite image processing (see Stolorz *et al.* (1995), for example).

Langley & Simon (1995) present an excellent overview of fielded applications of rule induction that comprise specific classification goals, and Brodley & Smyth (1997) outline a framework for the process of applying classifiers in real-world domains. Classification was also one of the data mining goals for the KDD '97 cup, a competition among data mining tool vendors at the *3rd International Conference of Knowledge Discovery and Data Mining* (Parsa, 1997).

Classification goals cover issues of the following subtasks:

- *Class Identification*

  If the data does not contain notions of classes in advance, the first subtask in classification is class identification. For example, clustering techniques separate data into groups of similar objects, and each group refers to a single class.

- *Class Description*

  If we are interested in concrete characteristics of objects that belong to the same class, we refer to this subtask in classification as class description. For example, conceptual clustering approaches solve class identification and class description subtasks in a single procedure. If we describe classes by an enumeration of specific objects, this description is *extensional*, whereas an explicit description in terms of characteristics yields an *intensional* explanation of classes.

- *Classifier Generation*

  If we want to use example data to generate a classifier which we then apply to assign classes to objects with unknown class information, we deal with the classification subtask classifier generation. For example, rule induction or top down induction of decision trees are typical approaches to solve the classifier generation subtask in classification.

- *Classifier Application*

  If we already possess a classifier (either generated in a previous step or revealed from any other source), we apply this classifier to new objects which do not include class assignments yet. For example, we use neural networks or decision trees in the classifier application subtask of classification goals.

In this dissertation, we focus on classification as a two step process:

1. Classifier Generation

2. Classifier Application

In the first step, data mining algorithms generate classifiers from *training sets*. A training set is a set of examples that include a characteristic description (usually an attribute-value description) plus the corresponding class label. A *classifier* normally uses information on the first $N - 1$ attributes (*predictive* attributes) to determine the class label (*predicted* attribute).[3] The second step is the application of generated classifiers to *test sets*. A test set is again a set of examples but for each example we only use the characteristic description, not the class label. The application task is now to use the generated classifier to predict the class label of an example in the test set based on its characteristic description. Usually, the training set and the test set are non-overlapping to ensure that we test the classifier on unseen instances.

In the following, we assume that the original data set comprises a classification task with pre-defined class labels. Hence, we do not need to identify classes first. The primary goal of classification is classifier generation and its application to unseen examples. We further restrict classification to qualitative classification goals, i.e., class labels are always symbolic, not numeric.

---

[3] We assume, the data set contains N attributes all in all.

2.4 Data Characteristics: Notations and Definitions

Data understanding is the next phase in KDD processes before data preparation that influences the definition of focusing tasks and the development or selection of appropriate focusing solutions. One of the main objectives in data understanding is the initial computation of data characteristics. Data mining goals and these data characteristics are two important aspects that guide focusing.

In this section, we define the terminology to describe database tables and specify basic data characteristics in terms of statistical values. The main purpose of these definitions is the introduction of notations which we use throughout the entire dissertation. We do not describe database technology and statistical concepts in detail but remind their basic intuitions. A fundamental understanding of these concepts is necessary to understand the specification of evaluation criteria for focusing success and the description of focusing solutions. For more detailed descriptions of databases and statistical aspects of data characteristics, we refer to Ullman (1982), Lockemann & Schmidt (1987), Hartung *et al.* (1987), Mason *et al.* (1989).

2.4.1 Database Tables

A (*relational*) *database* is a set of *database tables*. For simplicity, we only consider single tables. If information from different database tables is necessary to achieve the data mining goal in KDD, we assume that the integration of different sources as a result of the respective data preparation task is complete at this point. A single (database) table is a pair of a set of *attributes* and a set of *tuples* (see definition 2.4.2). Each *attribute* is characterized by its *name*, its specific *type*, and its domain of values. Each *tuple* contains a sequence of attribute values which specify concrete values for each attribute (see definition 2.4.1). In order to simplify some of the following definitions, we restrict an attribute domain to those values which occur in the (database) table.

The terminology for descriptions of data sets is not unique among all disciplines that contribute to KDD. Whereas the database community uses *tuple* to refer to a row in a database table, the same concept in statistics is an *observation*, *item* or *data point*, and in machine learning it is usually an *example*, *case*, or *instance*. Similarly, an *attribute* is also a *variable* or *feature*. We prefer to use tuples and attributes here, although we exceptionally use other terms as synonyms.

Definition 2.4.1 (Attributes and Tuples)

We define an attribute a_j *by a unique* name, *an attribute* type, *and a domain* $dom(a_j) = \{a_{j1}, a_{j2}, \ldots, a_{jk}, \ldots, a_{jN_j}\}$ *or* $dom(a_j) \subseteq \mathbb{R}$. N_j *denotes the number of (occurring) values in attribute domain* $dom(a_j)$.

Table 2.1: (Database) Table

	a_1	a_2	$\ldots$	a_j	$\ldots$	a_N
t_1	t_{11}	t_{12}	$\ldots$	t_{1j}	$\ldots$	t_{1N}
t_2	t_{21}	t_{22}	$\ldots$	t_{2j}	$\ldots$	t_{2N}
$\ldots$	$\ldots$	$\ldots$	$\ldots$	$\ldots$	$\ldots$	$\ldots$
t_i	t_{i1}	t_{i2}	$\ldots$	t_{ij}	$\ldots$	t_{iN}
$\ldots$	$\ldots$	$\ldots$	$\ldots$	$\ldots$	$\ldots$	$\ldots$
t_M	t_{M1}	t_{M1}	$\ldots$	t_{Mj}	$\ldots$	t_{MN}

We specify a tuple t_i as a sequence of values $t_i = (t_{i1}, t_{i2}, \ldots, t_{ij}, \ldots, t_{iN})$. Each value t_{ij} is an element of attribute domain $dom(a_j)$, i.e., $\forall t_{ij}, 1 \leq j \leq N : t_{ij} \in dom(a_j)$.

Definition 2.4.2 ((Database) Table)

Assume a set of attributes $A = \{a_1, a_2, \ldots, a_j, \ldots, a_N\}$, and a (multi-)set[4] of tuples $T = \{t_1, t_2, \ldots, t_i, \ldots, t_M\}$. We define a (database) table as a pair (A, T). N denotes the dimensionality of a database table, and M indicates its size.

We denote the power set, i.e., the set of all subsets, of A and T as $A^{\subseteq}$ and $T^{\subseteq}$, respectively.

Table 2.1 shows a (database) table and its index-driven notation. Index i always refers to a row in a (database) table and indicates the ith tuple. The second index j points to a column in a (database) table and depicts the jth attribute. Finally, we use a third index k to designate the kth value of an attribute domain.

For subsequent definitions, we introduce | and its different meanings. Definition 2.4.3 shows three different types of usage. First, we use | to indicate the size of a set, i.e., the number of elements in a set. Second, we denote the absolute value of a (possibly negative) number by surrounding |. Third, | also describes the beginning of conditions within sets and specifies subsets of objects which meet this condition.

Definition 2.4.3 (Notation: Usage of |)

Assume $X = \{x_1, \ldots, x_i, \ldots, x_M\}$, a number y, and a condition P.

(i) $| X |$ denotes the number of elements in X, i.e., $| X | := M$.

---

[4]This means, we allow identical tuples.

Table 2.2: (Database) Table Example

No.	Name	Age	Sex	Income	Married?	Credit
1	Miller	25	male	100000	no	good
2	Johnson	49	female	50000	yes	bad
3	Smith	55	female	65000	yes	bad
4	Smith	38	male	75000	no	bad
5	Henson	60	male	75000	yes	good
6	Carter	$\perp$	male	80000	yes	bad
7	Parker	45	female	40000	no	good
8	Ellison	67	$\perp$	$\perp$	no	good

(ii) $| y |$ *denotes the absolute value of* y, *i.e.,* $| y | := \begin{cases} y, & if \ y \geq 0 \\ -y, & if \ y < 0 \end{cases}$

(iii) $\{x_i \in X \mid P(x_i)\}$ *denotes the set of all elements in* X *that meet* P.

Example 2.4.1 ((Database) Table)

Table 2.2 shows an extraction of a simple database table in a credit scoring domain. This table contains eight tuples and six attributes. Each tuple includes information about a customer's name, age, sex, income, marital state, and whether the customer is a good or bad customer. The sixth and eighth tuples contain unknown values denoted as $\perp$.

If we are not interested in the entire information of a (database) table but want to consider only some attributes and their tuple values, we use *projections* of attributes and tuples. Definition 2.4.4 states projections and their notations for our purposes.

Definition 2.4.4 (Projections)

Assume a (database) table (A, T). *We define the* projection *of* A *and* T *to a subset of attributes as*

$$A_{[j; \, j']} \ := \ \{a_j, a_{j+1}, \dots, a_{j'-1}, a_{j'}\}, \text{ and}$$

$$T_{[j; \, j']} \ := \ \{(t_{ij}, t_{i(j+1)}, \dots, t_{i(j'-1)}, t_{ij'}) \mid t_i \in T\}, \text{ respectively.}$$

Similarly, we define the projection of a tuple $t_i \in T$ *as*

$$t_{i[j;\,j']} \quad := \quad (t_{ij}, t_{i(j+1)}, \ldots, t_{i(j'-1)}, t_{ij'}).$$

The basic definitions of (database) tables and their projections determine notations for descriptions of tables. However, tables differ in their specific characteristics. Information about some fundamental characteristics of (database) tables is usually obtained from *data dictionaries*. For example, data dictionaries contain information about attribute names as well as attribute types and domains.

Attribute types give information on values in attribute domains. We distinguish between *qualitative* and *quantitative* attributes (see definition 2.4.5). Domains of qualitative attributes contain a set of different values without any relations among them. Domains of quantitative attributes are a subset of real numbers.[5]

Definition 2.4.5 (Attribute Types)

Assume a (database) table (A, T), *and an attribute* $a_j \in A$.

a_j *is* qualitative, *if* $dom(a_j) = \{a_{j1}, a_{j2}, \ldots, a_{jk}, \ldots, a_{jN_j}\}$.

a_j *is* quantitative, *if* $dom(a_j) \subseteq \mathbb{R}$.

Note, although we do not further restrict values of quantitative attributes and each value is any number in the specified superset, we always denote by $dom(a_j)$ the finite set of occurring values for attribute a_j in the (database) table. This assumption simplifies some subsequent definitions, and we ensure that at least one tuple in the database represents a particular value in $dom(a_j)$.

Example 2.4.2 (Attribute Types)

In table 2.2, name, sex, marital state, and credit are qualitative attributes whereas age and income are quantitative attributes.

Often, we also compute more interesting characteristics in terms of statistics from information in a database and its data dictionary. For example, SQL queries help to compute such information. Definition 2.4.6 introduces notations for *attribute value frequencies* which count the number of tuples in a (database) table with a specific value of an attribute domain. We also define *joint* frequencies which consider more than single attributes and their values simultaneously.

---

[5]Note, in the following, we only use the characteristic of quantitative attributes that there exists an order relation on its domain and that there is a notion of distance between values in its domain. However, such attributes are usually numeric, i.e., either a set of integers, fractions, or real values.

Definition 2.4.6 (Attribute Value Frequencies)

Assume a (database) table (A, T), a set of tuples $T' \subseteq T$, an attribute $a_j \in A$, and an attribute value $a_{jk} \in dom(a_j)$. We define the absolute frequency of a_{jk} within T' as a function

$$n_{jk} : T^{\subseteq} \longrightarrow [0; \ M],$$

$$n_{jk}(T') := | \ \{t_i \in T' \ | \ t_{ij} \equiv a_{jk}\} \ |.$$

Similarly, we define the joint absolute frequency *of attribute values a_{1k_1}, ..., a_{jk_j}, ..., a_{Nk_N}, $a_{1k_1} \in dom(a_1)$, ..., $a_{jk_j} \in dom(a_j)$, ..., $a_{Nk_N} \in dom(a_N)$, within T' as a function*

$$n_{k_1 \dots k_j \dots k_N} : T^{\subseteq} \longrightarrow [0; \ M],$$

$$n_{k_1 \dots k_j \dots k_N}(T') := | \ \{t_i \in T' \ | \ t_{ij} \equiv a_{jk_j} \ \forall j, \ 1 \leq j \leq N\} \ |.$$

2.4.2 Statistical Values

From a statistics point of view, a database represents a *population*, and each attribute in a (database) table defines a (random) *variable*. In this sense, the goal of focusing is drawing *samples* from a population. Note, if we keep terminology consistent with statistics, a (database) table is already a sample of a population which is defined by the cartesian product of all attribute domains. Since a database usually does not contain each possible combination of different values, a database only represents a sample of all possible tuples and hence is a sample of a larger population. However, we regard the existing set of tuples in a (database) table as the population and the focusing output as the sample.

Meta-Data

In statistics, we often analyze statistical values such as *modus, mean, range, variance,* and *standard deviation*. These statistical values describe *meta-data* on the population or samples, respectively. Meta-data is information which is not explicitly part of the data but is inherently specified by the data, and we are able to compute meta-data with mathematical mechanisms. This information is only valuable, if we know the origin of its computation. For example, the mean of an attribute is only useful, if we know which attribute we used for computation and what its meaning is. In some sense, we consider computations of statistical values also as specific types of data mining, since meta-data abstracts from the original data and represents information which is not explicitly in the data.

The following definitions specify statistical values which we use to analyze success of focusing solutions. We test the coincidence between statistical values in the (database) table and the corresponding value in the output of focusing solutions. The level of agreement between these values indicates the representativeness of samples. The better statistical values in the sample match the same values in the population, the higher is the representativeness of the sample. Recall, we define statistical values in order to introduce their notations, and we do not describe their meanings and characteristics in detail.

In order to distinguish between true statistical values and their realizations in samples, we explicitly indicate the set of tuples which is used to compute the value as an additional parameter in each definition. Note, we do not use abbreviations such as in statistics which usually omit this additional parameter, if its value corresponds to the entire population.

Definition 2.4.7 (Modus)

Assume a (database) table (A, T), *a set of tuples* $T' \subseteq T$, *and an attribute* $a_j \in A$. *We define the* modus *of* a_j *within* T' *as a function*

$$m_j : T^{\subseteq} \longrightarrow dom(a_j), \quad m_j(T') := a_{jk} \text{ with } n_{jk}(T') \equiv \max_{k'=1}^{N_j}\{n_{jk'}(T')\}.$$

Definition 2.4.8 (Mean)

Assume a (database) table (A, T), *a set of tuples* $T' \subseteq T$, *and a quantitative attribute* $a_j \in A$. *We define the* mean *of* a_j *within* T' *as a function*

$$\mu_j : T^{\subseteq} \longrightarrow \,]-\infty;\, \infty[, \quad \mu_j(T') := \frac{1}{|T'|} \cdot \sum_{i=1}^{|T'|} t_{ij}.$$

Definition 2.4.9 (Range)

Assume a (database) table (A, T), *a set of tuples* $T' \subseteq T$, *and a quantitative attribute* $a_j \in A$. *We define the* range *of* a_j *within* T' *as a function*

$$r_j : T^{\subseteq} \longrightarrow [0;\, \infty[\,, \quad r_j(T') := |\max_{i=1}^{|T'|}\{t_{ij}\} - \min_{i=1}^{|T'|}\{t_{ij}\}|.$$

Definition 2.4.10 (Variance and Standard Deviation)

Assume a (database) table (A, T), *a set of tuples* $T' \subseteq T$, *and a quantitative attribute* $a_j \in A$. *We define the* variance *of* a_j *within* T' *as a function*

$$\sigma_j^2 : T^{\subseteq} \longrightarrow [0;\, \infty[, \quad \sigma_j^2(T') := \frac{1}{|T'|-1} \cdot \sum_{i=1}^{|T'|}(t_{ij} - \mu_j(T'))^2.$$

We define the standard deviation *of* a_j *within* T' *as a function*

$$\sigma_j : T^{\subseteq} \longrightarrow [0; \infty[,$$

$$\sigma_j(T') := \sqrt{\sigma_j^2(T')} = \sqrt{\frac{1}{|T'|-1} \cdot \sum_{i=1}^{|T'|} (t_{ij} - \mu_j(T'))^2}.$$

Note, the finite error correction $\frac{1}{|T'|-1}$ in definition 2.4.10 ensures that σ_j^2 and σ_j are unbiased estimators of the true values for σ^2 and σ in the population. Recall, each subset of tuples (including the entire set of tuples) is only a sample of a larger population.

Redundancy

In real-world domains, we often observe high redundancy in (database) tables (Moller, 1993). A tuple is *redundant*, if an additional identical tuple in the same table exists. Redundant tuples do not contribute new information except that they increment the number of tuples with identical characteristics. We define the *redundancy (factor)* of sets of tuples as 1 minus the fraction of the number of tuples in the (database) table after removal of all duplicates and the number of tuples in the original (database) table (see definition 2.4.11).

Definition 2.4.11 (Redundancy (Factor))

Assume a (database) table (A, T), a set of tuples $T' \subseteq T$, and $T'_{\{\}} := \{t_i \in T' \mid \forall i', 1 \leq i' \leq i - 1 : t_{i'} \neq t_i\}$. We define the redundancy (factor) within T' as a function

$$R : T^{\subseteq} \longrightarrow [0; 1[, \quad R(T') := 1 - \frac{|T'_{\{\}}|}{|T'|}.$$

2.5 Data Mining Algorithms

In the previous two sections, we considered aspects that influence the definition of focusing tasks and the development or selection of appropriate focusing solutions in KDD phases before data preparation. Now, we step forward in the process and turn to the data mining phase. Note, although we do not consider data exploration, focusing is also a meaningful preparation task for this phase as well. In the following, we assume that we achieve data mining goals by applications of data mining algorithms. This assumption simplifies the complexity of KDD processes.

Recall, we already selected classification as the primary data mining goal in this dissertation. Hence, we focus aspects of data mining algorithms to classification algorithms. In this section, we first introduce notations for the classifier generation and classifier application steps in classification. Then, we describe basic properties of two data mining algorithms for classification goals in more detail.[6]

2.5.1   Classification Algorithms

Recall, we consider classification as a two step process consisting of classifier generation and classifier application. In this section, we specify both steps in more detail and introduce notations for training and test set as well as classifiers. Training and test set are non-overlapping subsets of a subset of tuples, and classifiers are functions that map characteristic descriptions of tuples in terms of attribute values for the first $N - 1$ attributes to class labels which refer to values of the last attribute (see definition 2.5.1).

Definition 2.5.1 (Classification)

Assume a (database) table (A, T), a set of tuples $T' \subseteq T$, a training set $T_{train} \subseteq T'$, a test set $T_{test} \subseteq T'$ with $T' = T_{train} \cup T_{test}$ and $T_{train} \cap T_{test} = \emptyset$, and a qualitative attribute $a_N \in A$ with domain $dom(a_N) = \{a_{N1}, \ldots, a_{Nk}, \ldots, a_{NN_N}\}$. We define classification as a two-step process:

1.   Classifier Generation

     Generate a classifier $\phi : T_{[1;\ N-1]} \longrightarrow dom(a_N)$ using T_{train}.

2.   Classifier Application

     Apply ϕ to T_{test}.

We call a_N the class attribute and $dom(a_N)$ the set of class labels.

In the classifier generation step, we apply a data mining algorithm to the training set in order to build a classifier. In the classifier application step, we use this classifier to assign class labels in the domain of the predicted attribute to each of the tuples in the test set. In statistics and machine learning exist numerous algorithms to generate classifiers. For example, various methods of *discriminant analysis* or *regression* approaches, *rule induction* algorithms such as CN2 and ITrule, *decision tree* procedures such as C4.5 and NewID, neural networks, and *nearest neighbor* mechanisms build different types of classifiers. We refer to Michie *et al.* (1994) for an overview and description of these and other classification algorithms.

In this dissertation, we consider two specific classification algorithms, *top down induction of decision trees* and *nearest neighbor classifiers*, which represent two different learning paradigms. We select those algorithms since top down induction of decision trees is widely accepted as one of the best approaches to classification and often applied to solve business questions in industry, and be-

---

[6]In a strong sense, we also distinguish between an algorithm and its (various) implementation(s). In this section, we focus on algorithms rather than implementations.

cause nearest neighbor classifiers are generally slow and need focusing approaches in order to scale up to more than only a few hundred tuples. Consequently, many applications will benefit from focusing solutions that enable more appropriate usage of top down induction of decision trees and nearest neighbor classifiers.

An additional important reason for the choice of top down induction of decision trees and nearest neighbor classifiers for further examinations is their ability to handle all types of data characteristics including mixtures of qualitative and quantitative attributes, missing values, noise, and other typical characteristics of real-world data sets. The only restriction in case of top down induction of decision trees is the limitation for the class attribute, which has to be a qualitative attribute.

2.5.2   Top Down Induction of Decision Trees

Decision trees are favorite classifiers in KDD since the construction of decision trees is efficient, and their classification accuracy is often higher than the classification accuracy of competitive algorithms. Although we assume that the reader is familiar with top down induction of decision trees, we briefly describe its main characteristics. First, we outline the structure of decision trees, and then we present a short description of the classifier generation step.

Decision Trees

In top down induction of decision trees, the classifier is a decision tree. Each node in decision trees corresponds to a test on a specific attribute and its values. Each test results in branches which represent different outcomes of the test. For qualitative attributes, the set of outcomes is the set of different values in the respective attribute domain. For quantitative attributes, top down induction of decision trees specifies a threshold value and generates two separate branches. The first branch covers all tuples with values below or equal to this threshold, and the second branch includes tuples with values above this threshold. For each branch, decision trees contain subtrees of the same structure. Terminal nodes in decision trees cover only tuples of the same class or small subsets of tuples for which it is no longer worth to proceed with.

Example 2.5.1 (Decision Tree)

Figure 2.6 shows the decision tree that results from the application of top down induction of decision trees to the example database in a credit scoring domain (see table 2.2, page 27). In this example, the first test attribute is age. If the age value of a customer is larger than 55, this customer is a good customer. Otherwise, the decision tree proceeds testing of attribute values with attribute married?. If the customer is married, the customer is a bad customer. If the customer is not married, we need to test the age of a customer again.

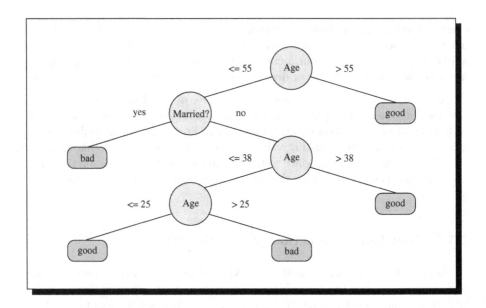

Figure 2.6: Decision Tree Example

Example 2.5.1 already illustrates some characteristics of decision trees. For instance, it is possible that quantitative attributes are the test attribute more than once at different tree levels. Figure 2.6 also shows the restricted representation of decision trees. For quantitative attributes, it is only possible to generate two branches according to a single threshold value. For example, if we were able to use more than two intervals rather than only a single threshold value, the age nodes at lower levels were obsolete.

Algorithm TDIDT

The classifier generation step in top down induction of decision trees uses a divide-and-conquer strategy to build decision trees. The algorithm starts with the entire set of tuples in the training set, selects the *best* attribute which conveys maximum information for classification, and generates a test node for this attribute. Then, top down induction of decision trees divides the current set of tuples according to their values of the current test attribute. It continues with each subset of tuples with the same value (or values in the same range of values) in the same way. Classifier generation stops, if all tuples in a subset belong to the same class, or if it is not worth to proceed with an additional separation into further subsets. It is not worth to proceed, if further attribute tests yield only information for classification below a pre-specified threshold.

In the classifier application step, we apply the resulting decision tree to classify tuples. As the example tree (see figure 2.6) indicates, decision trees propagate tuples down the tree according to the outcome of each test attribute. At terminal nodes, the application of decision trees assigns the class label (or the majority of class labels) at this node to the new tuple.

Entropy

The essential step in top down induction of decision trees is the selection of the *best* test attribute. This classification algorithm uses an *information theoretic* evaluation measure for the selection of the best attribute. Since we reuse this measure to automatically compute attribute relevance later, we briefly describe this measure.

The idea is to measure information conveyed by splits on specific attributes. For subsets of tuples, this information depends on probability distributions of attribute values in these sets. If we randomly select a tuple, we estimate the probability that this tuple belongs to a specific class by computing the frequency of tuples with the same class label divided by the total number of tuples. The information of the corresponding message that a randomly selected tuple belongs to a specific class is then the negative logarithmus dualis of this relative frequency. If we are interested in the entire expected information pertaining class membership, we sum up information for each class. This quantity is known as the expected average information needed to identify the class of a tuple in the given set. We refer to this information as the *entropy* of a set (see definition 2.5.2).

Definition 2.5.2 (Entropy)

Assume a (database) table (A, T), a set of tuples $T' \subseteq T$, and a class attribute $a_N \in A$. We define the entropy *of T' as a function*

$$H : T^{\subseteq} \longrightarrow [0; \infty[, \quad H(T') := -\sum_{k=1}^{N_N} \frac{n_{Nk}(T')}{|T'|} \cdot \log_2 \left(\frac{n_{Nk}(T')}{|T'|} \right).$$

The next step towards selection of the best test attribute is an extension of entropy to more than single sets of tuples. In order to measure the information of an attribute split and the resulting partition of a set of tuples into subsets according to attribute values for a specific attribute, we cumulate the weighted information of each subset. This results in the definition of the *entropy of an attribute split* (see definition 2.5.3).

Definition 2.5.3 (Entropy of an Attribute Split)

Assume a (database) table (A, T), a set of tuples $T' \subseteq T$, a qualitative attribute $a_j \in A$, and $T'_{jk} := \{t_i \in T' \mid t_{ij} \equiv a_{jk}\}$. We define the entropy of an attribute *split on a_j as a function*

$$H_j : T^{\subseteq} \longrightarrow [0; \infty[, \quad H_j(T') := \sum_{k=1}^{N_j} \frac{n_{jk}(T')}{|T'|} \cdot H(T'_{jk}).$$

For quantitative attributes, top down induction of decision trees separates the set of occurring values into two intervals by specifying a threshold value. In order to find the best threshold, the algorithm checks all means between two different occurring values. The test threshold is then set to the highest occurring value below the mean which results in maximum entropy. This strategy ensures that selected thresholds always correspond to occurring values in the set of tuples.

Information Gain

The next definition specifies the *information* that we *gain* if we split on an attribute. Therefore, we compute the information that we need to classify a tuple correctly without and with a split on this attribute. The first information without attribute split is the entropy (see definition 2.5.2), and the second information corresponds to the entropy of an attribute split (see definition 2.5.3). The difference between these two information values is the information that we gain by an attribute split (see definition 2.5.4).

Definition 2.5.4 (Information Gain)

Assume a (database) table (A, T), a set of tuples $T' \subseteq T$, and an attribute $a_j \in A$. We define the information gain of an attribute split on a_j as a function

$$H_j^+ : T^{\subseteq} \longrightarrow [0; \infty[, \quad H_j^+(T') := H(T') - H_j(T').$$

First implementations of top down induction of decision trees such as ID3 (Quinlan, 1986) use this information gain measure to select the best split attribute. However, experiments with information gain show that this criterion tends to prefer attributes with many values in its domain. For this reason, Quinlan (1993) proposes a normalization of information gain to avoid this tendency. The resulting evaluation measure is called *information gain ratio* (see definition 2.5.5).

Definition 2.5.5 (Information Gain Ratio)

Assume a (database) table (A, T), a set of tuples $T' \subseteq T$, and an attribute $a_j \in A$. We define the information gain ratio of an attribute split on a_j as a function

$$I_j^+ : T^{\subseteq} \longrightarrow [0; \infty[, \quad I_j^+(T') := \frac{H_j^+(T')}{-\sum_{k=1}^{N_j} \frac{n_{jk}(T')}{|T'|} \cdot \log_2 \left(\frac{n_{jk}(T')}{|T'|} \right)}.$$

2.5.3 Nearest Neighbor Classifiers

We consider nearest neighbor classifiers as the second type of classification algorithms in this dissertation. Similar to top down induction of decision trees, we assume that the reader is generally familiar with nearest neighbor classifiers. Since we return to nearest neighbor approaches and reuse similarity measures later, we outline a basic nearest neighbor classification algorithm and specify examples of similarity measures subsequently.

Algorithm NN

The classifier generation step in basic nearest neighbor algorithms is quite simple. They store all existing tuples in the training set and select a similarity measure (or a distance function, respectively). Classifier application to new tuples in the test set is then the retrieval of the most similar tuple in the training set according to the given similarity measure (or the distance function, respectively). Nearest neighbor classifiers return the class label of this tuple as the predicted class label of an unseen tuple in the test set.

Algorithm 2.5.1 $NN(T, t, Sim)$

---

begin
  $i := 1$;
  $i^* := 1$;
  $sim^* := Sim(t_{i^*}, t)$;
  while $i \leq |T|$ **do**
      $i := i + 1$;
      $sim := Sim(t_i, t)$;
      if $sim > sim^*$
          then $i^* := i$;
                $sim^* := sim$;
      endif
  enddo
  return$(t_{i^* N})$;
end;

Algorithm 2.5.1 outlines a nearest neighbor classifier. This algorithm uses a set of tuples T and a similarity measure Sim to assign a class label to a given tuple t. Therefore, NN computes the similarity between each tuple t_i in T and tuple t. If the similarity between the current tuple in T and t is higher than the previous maximum similarity between a tuple in T and tuple t, NN adapts similarity sim^* to this value. After consideration of each tuple in T, sim^* holds the maximum similarity between a tuple in T and tuple t, and index i^* refers to the most similar tuple in comparison to t. This tuple t_{i^*} is the *nearest neighbor* of t in T. The nearest neighbor classifier returns the class label of this nearest neighbor as the predicted class label of the given tuple t.

Note, whereas algorithm TDIDT in the previous section generates a classifier, this algorithm refers to classifier application. The classifier ϕ in case of nearest neighbor algorithms is a set of tuples plus a similarity measure (or a distance function, respectively).

Enhancements to NN

An extension of basic nearest neighbor classifiers uses k nearest neighbors instead of only *the* nearest neighbor to determine the class label of an unseen tuple in the test set. The set of the k most similar tuples constitutes the set of k nearest neighbors. A voting mechanism among all class labels of these k nearest neighbors is then used to compute the final predicted class label for a tuple in the test set. For example, a simple voting mechanism returns the most often occurring class label within the set of k nearest neighbors as the predicted class label of an unseen tuple. More advanced techniques weigh the vote of each nearest neighbor (e.g., Cost & Salzberg, 1993). For instance, each vote is weighted according to the similarity between the nearest neighbor and the unseen tuple. The higher the similarity is, the higher is the weight of the respective neighbor's vote.

Another line of extension in nearest neighbor classifiers concerns the number of tuples in the training set retained for classification. For example, *edited nearest neighbor* approaches generate a more condensed representation of the set of tuples T in advance (Chang, 1974; Dasarathy, 1991), and some *instance-based learning* algorithms reduce storage requirements of nearest neighbor classifiers by selecting a subset of all tuples in T before the retrieval of nearest neighbors (Aha *et al.*, 1991). Since these approaches also contribute to focusing solutions, we consider them in more detail in section 4.1.4.2.

Similarity Measures

Since nearest neighbor classifiers mainly rely on an appropriate notion of similarity, we discuss some specific approaches at this point. We emphasize on *similarity measures* rather than other mechanisms to define notions of similarity. We also reuse similarity measures in the development of more intelligent focusing solutions later.

In general, we define similarity measures as reflexive, symmetric mappings between the cartesian product of a set of objects and interval [0; 1] (see definition 2.5.6). *Reflexivity* means that each object has maximum similarity to itself, whereas *symmetry* ensures that the similarity between two objects is independent of the direction of comparison. In general, we can only assume reflexivity as a meaningful requirement for similarity measures, since for any set of objects, an object has maximum similarity to itself. We regard symmetry as an additional reasonable characteristic of similarity in the context of (database) tables, although this assumption is not valid in all domains.

Definition 2.5.6 (Similarity Measure)

Assume a set of objects X. We define a similarity measure as a function

$$sim : X \times X \longrightarrow [0; 1] \; with$$

(i) $\forall x \in X : \quad sim(x, x) = 1$ (reflexivity),

(ii) $\forall x, y \in X : \quad sim(x, y) = sim(y, x)$ (symmetry).

Definition 2.5.6 only restricts similarity measures to all possible reflexive and symmetric mappings between pairs of objects and real values between 0 and 1. Now, we define specific similarity measures in order to compute similarities between tuples of (database) tables. Since tuples are sequences of attribute values, we first specify *local* similarity measures on attribute domains, and then cumulate these measures to overall similarities between tuples.

Note, subsequent definitions of similarity measures only specify examples. For each type of similarity measure, many alternative suggestions exist (see Bock (1974), Tversky (1977), Datta (1997), Wilson & Martinez (1997b), for example). Since similarity measures are not the main focus of the work presented here, we only describe some measures which are efficient in their computation and which we reuse in this dissertation.

Local Similarity Measures

For local similarity measures, X in definition 2.5.6 is an attribute domain of some attribute in a (database) table. The specific type of local similarity depends on the attribute type. The more information the attribute domain contains, the more fine-grained local similarity measures are able to reflect the underlying nearness of different values.

Purely qualitative attribute domains without any order relations among its values contain the minimum possible information. The respective local similarity measure only compares two values and checks whether they are identical or not. The similarity is minimal, if two values do not coincide, and it is maximal, if both values are the same (see definition 2.5.7).

Definition 2.5.7 (Local Similarity for Qualitative Attributes)

Assume a (database) table (A, T), and a qualitative attribute $a_j \in A$. We define a local similarity measure for a_j as a function

$$sim_j : dom(a_j) \times dom(a_j) \longrightarrow \{0, 1\},$$

$$sim_j(a_{jk}, a_{jk'}) := \begin{cases} 1, & if \; a_{jk} \equiv a_{jk'} \\ 0, & else \end{cases}$$

If an attribute is quantitative rather than qualitative, its domain contains additional information on the order of attribute values and an induced notion of distance between attribute values. Since quantitative attribute domains are subsets of real values, we utilize the absolute difference between two values as a notion of distance and define local similarities for quantitative attributes as the corresponding similarity measure (see definition 2.5.8).

Definition 2.5.8 (Local Similarity for Quantitative Attributes)

Assume a (database) table (A, T), and a quantitative attribute $a_j \in A$. We define a local similarity measure for a_j as a function

$$sim_j : dom(a_j) \times dom(a_j) \longrightarrow [0; 1],$$

$$sim_j(a_{jk}, a_{jk'}) := 1 - \frac{\mid a_{jk} - a_{jk'} \mid}{r_j(T)}.$$

Note, the normalization by a division of the distance between two values and the range (see definition 2.4.9, page 30) is necessary to ensure that the local similarity measure for quantitative attributes meets the requirements in the general definition of similarity measures (see definition 2.5.6, page 39). Since each attribute domain is finite, minimum and maximum occurring values always exist.

Cumulated Similarity Measures

Now, we cumulate local similarities to overall similarities between tuples. A simple cumulation is the sum of local similarities between values for each attribute and its normalization by dividing this sum and the number of attributes. Since attributes in (database) tables have different relevance, we suggest to add *relevance weights* for each attribute. For the moment, we assume that attribute relevance weights are given in advance. In section 4.2.3.4, we consider different methods to determine attribute weights that reflect the relevance of attributes. For discussions of attribute relevance in nearest neighbor classifiers, see Wettschereck (1994), Wettschereck *et al.* (1995), Kohavi *et al.* (1997), for example.

Definition 2.5.9 (Weighted Cumulated Similarity)

Assume a (database) table (A, T), local similarity measures sim_j for attribute a_j, $1 \leq j \leq N$, relevance weights $w_j \geq 0$ for attribute a_j, $1 \leq j \leq N$, and $w := \sum_{j=1}^{N} w_j$. We define a weighted cumulated similarity measure as a function

$$Sim : T \times T \longrightarrow [0; 1], \quad Sim(t_i, t_{i'}) := \frac{1}{w} \cdot \sum_{j=1}^{N} w_j \cdot sim_j(t_{ij}, t_{i'j}).$$

Definition 2.5.9 specifies a weighted cumulated similarity measure. We define the overall similarity between tuples in (database) tables as the weighted sum of local similarities for each attribute. Each weight corresponds to attribute relevance and is a positive real value. In order to meet the general requirements of similarity measures (see definition 2.5.6, page 39), we normalize the weighted sum by dividing by the sum of all attribute relevance weights. In this definition, subscript j indicates that the local similarity measure sim_j must have the correct type according to the attribute type of attribute a_j.

Properties of Weighted Cumulated Similarity Measures

Since we reuse similarity measures at several points in this dissertation, we point out two useful properties of weighted cumulated similarities.

Proposition 2.5.1 (Maximum Similarity)

Assume a (database) table (A, T), a weighted cumulated similarity measure Sim, and $w_j \neq 0$, $1 \leq j \leq N$.

$$\forall t_i, t_{i'} \in T :$$

$$Sim(t_i, t_{i'}) = 1 \quad \Longleftrightarrow \quad t_i \equiv t_{i'}, i.e., \forall j, \ 1 \leq j \leq N : t_{ij} = t_{i'j}$$

Proof:

This property follows from definitions 2.5.7, 2.5.8, and 2.5.9. □

Proposition 2.5.1 states that the weighted cumulated similarity between tuples is 1, if and only if all attribute values are identical. This property ensures that the similarity between tuples only reveals maximum values, if the tuples are completely identical. Note, if some attribute relevance weights are zero, it is sufficient for maximum similarity between tuples, if two tuples coincide for all attributes with non-zero weights.

Proposition 2.5.2 (Δ-Inequality)

Assume a (database) table (A, T), and a weighted cumulated similarity measure Sim.

$$\forall t_i, t_{i'}, t_{i''} \in T : \quad 1 - Sim(t_i, t_{i'}) + 1 - Sim(t_{i'}, t_{i''}) \geq 1 - Sim(t_i, t_{i''})$$

Proof:

See appendix C, proposition C.0.1, page 277. □

Proposition 2.5.2 shows that the triangle inequality holds for the distance function which corresponds to the weighted cumulated similarity measure specified in definition 2.5.9. This means, this similarity measure corresponds to a metric distance function.

Similarity Measures and Unknown Values

Since we are dealing with real-world data sets and we do not assume complete data without any unknown values, we have to define local similarities in case of comparisons with unknown values. We distinguish the following strategies to handle unknown values in similarity measures (see definition 2.5.10). The first alternative proposes to use a *default strategy*. In this case, we compare an attribute value and an unknown value as if the unknown value corresponds to a specific default value in the respective attribute domain.[7] This strategy is simple but does not make any difference in similarity values for different values in comparison to unknown values.

Definition 2.5.10 (Local Similarity with Unknown Values)

Assume a (database) table (A, T), an attribute $a_j \in A$, a local similarity measure sim_j, an attribute value $a_{jk} \in dom(a_j)$, and an unknown value $\perp$). We propose the following strategies to compute local similarities with unknown values:

1. *Default Strategy*

   $sim_j(a_{jk}, \perp) := sim_j(a_{jk}, a_{j\perp})$ *for a default value* $a_{j\perp} \in dom(a_j)$

2. *Optimistic Strategy*

   $$sim_j(a_{jk}, \perp) := \max_{k'=1, k' \neq k}^{N_j} \{sim_j(a_{jk}, a_{jk'})\}$$

3. *Pessimistic Strategy*

   $$sim_j(a_{jk}, \perp) := \min_{k'=1, k' \neq k}^{N_j} \{sim_j(a_{jk}, a_{jk'})\}$$

4. *Average Strategy*

   $$sim_j(a_{jk}, \perp) := \frac{1}{N_j - 1} \cdot \sum_{k'=1, k' \neq k}^{N_j} sim_j(a_{jk}, a_{jk'})$$

5. *Ignore Strategy*

   $w_j := 0$

The second, third, and fourth alternatives use the same idea to compute local similarities between attribute values and unknown values. Whereas the *optimistic strategy* assumes maximum similarity among all occurring similarity values as the value for the comparison between an attribute value and an unknown value, the *pessimistic strategy* computes the minimum similarity among occurring values as a representation of the similarity between an attribute value and an unknown value. Both approaches assume extreme values for similarities in case of unknown values, and hence both approaches eventually refer to an

---

[7]For example, *maximum*, *minimum*, *modus*, or *mean* values in $dom(a_j)$ are candidates for this default value.

Table 2.3: Similarity Matrix Example

No.	1	2	3	4	5	6	7	8
1	1.00	0.09	0.10	0.47	0.39	0.28	0.36	0.40
2	0.09	1.00	0.66	0.33	0.33	0.42	0.39	0.11
3	0.10	0.66	1.00	0.49	0.39	0.46	0.34	0.14
4	0.47	0.33	0.49	1.00	0.35	0.49	0.32	0.26
5	0.39	0.33	0.39	0.35	1.00	0.49	0.29	0.37
6	0.28	0.42	0.46	0.49	0.49	1.00	0.06	0.00
7	0.36	0.39	0.34	0.32	0.29	0.06	1.00	0.50
8	0.40	0.11	0.14	0.26	0.37	0.00	0.50	1.00

extremely inappropriate value. The *average strategy* is a compromise between both extreme value approaches and computes the similarity between an attribute value and an unknown value as the average value of similarities among occurring values in the respective attribute domain. Note, all three alternatives take into account specific attribute values and yield different similarities for different values.

Since the default strategy is likely to yield an inappropriate local similarity, and the optimistic, pessimistic, and average strategies rely on expensive computations, we finally propose to neglect contributions of local similarities with unknown values. The *ignore strategy* sets the respective attribute relevance weight to zero, if an unknown value occurs, and consequently this strategy does not consider this attribute to compute the overall weighted cumulated similarity between two tuples. We use this strategy in case of unknown values in the implementations of more intelligent sampling techniques.

Example 2.5.2 (Nearest Neighbor Classifier)

Table 2.3 depicts all weighted cumulated similarities between pairs of different tuples in the example database in a credit scoring domain (see example 2.2, page 27), if we assume that each attribute relevance weight is 1.0.

If the training set contains tuples 1 to 6 and the test set includes tuples 7 and 8, algorithm NN classifies tuple 7 as bad and tuple 8 as good, since tuple 2 is the nearest neighbor of tuple 7 and tuple 1 is the nearest neighbor of tuple 8 in the training set.

2.6   Selecting the Focusing Context

In summary, this chapter provides the overall focusing context in KDD which influences the definition of focusing tasks and the development or selection of appropriate focusing solutions. First, we analyzed the KDD process in more detail and outlined a process model for KDD projects. We identified three aspects in business understanding, data understanding, and data mining phases which have the most relevant impact on focusing as a data preparation task. These aspects include data mining goals, data characteristics, and data mining algorithms.

Then, we considered each of these aspects separately. We briefly described possible data mining goals and selected classification as the primary data mining goal in this dissertation. We also introduced notations and concepts for database tables and statistical values to enable an analysis of data characteristics. Finally, we characterized classification algorithms and chose two specific approaches for further examination. In this dissertation, we apply top down induction of decision trees and nearest neighbor classifiers.

The following list consolidates the first specialization of the scope of work presented here:

- *Data Mining Goal*

  The data mining goal is classification.

- *Data Characteristics*

  We do not restrict data characteristics except that the set of attributes contains a qualitative class attribute.

- *Data Mining Algorithms*

  The data mining algorithm is either top down induction of decision trees or nearest neighbor.

In summary, we concentrate the development, analysis, and evaluation of focusing solutions for data mining on the data mining goal classification and on the data mining algorithms top down induction of decision trees and nearest neighbor classifiers.

Chapter 3

Focusing Tasks

In this chapter, we specify a framework for the definition of focusing tasks. We describe the concept of focusing specifications and outline relations between focusing context and focusing tasks as well as their solutions. We also define evaluation criteria for focusing success. The primary goal of this chapter is the definition and selection of specific focusing tasks for further examination.

3.1 Focusing Concepts: An Overview

In the following sections, we define *focusing concepts* in detail. In order to ensure an initial understanding of each concept, we outline basic terminology and relations between focusing concepts at the beginning of this chapter. Figure 3.1 presents an overview of these concepts and their relations. We distinguish between the following concepts:

- *Focusing Task*

  A focusing task consists of focusing specification and focusing context.

- *Focusing Specification*

  A focusing specification consists of *focusing input, focusing output*, and *focusing criterion*.

- *Focusing Context*

  A focusing context consists of *data mining goal, data characteristics*, and *data mining algorithm*.

- *Focusing Solution*

  A focusing solution solves a focusing task. It takes the focusing input and generates a focusing output within the focusing context. It attempts

T. Reinartz: Focusing Solutions for Data Mining, LNAI 1623, pp. 45-84, 1999
© Springer-Verlag Berlin Heidelberg 1999

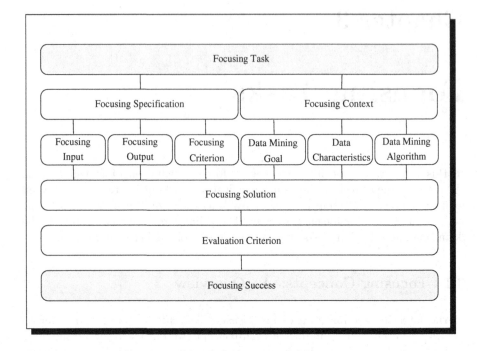

Figure 3.1: Focusing Concepts

to generate a focusing output that meets (or optimizes) the evaluation criterion.

Note, we distinguish between *deterministic* focusing solutions that always generate the same focusing output for specific parameter settings, and *nondeterministic* focusing solutions that probably generate different focusing outputs for the same parameter setting.

- *Evaluation Criterion*

  An evaluation criterion specializes the general notion of the focusing criterion, and we consider evaluation criteria as specific realizations of focusing criteria. Focusing specification and focusing context determine the evaluation criterion which is a measure of *focusing success* for focusing solutions.

Subsequently, we introduce precise definitions of focusing concepts and use these definitions in order to specify focusing tasks in detail. The definition of focusing tasks refines the general goal of focusing and its intuitive meaning. Focusing tasks translate the intuitive meaning into formal definitions, and these formalizations specify requirements on systems which solve focusing tasks. Focusing tasks include the focusing context which determines the primary purpose of focusing in KDD.

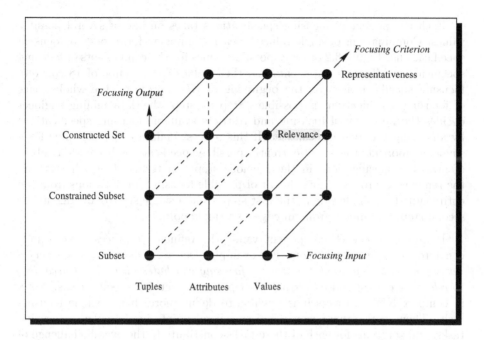

Figure 3.2: Focusing Specifications

3.2 Focusing Specification

In general, focusing concerns data reduction in order to enable the application of data mining algorithms or to improve efficiency of data mining. In some cases, focusing also leads to better data mining results. At more detailed levels, focusing attempts to solve more specific tasks. Data reduction is possible in different ways and for different purposes. In this section, we describe these different ways and different purposes.

We define the concept of *focusing specifications* that characterize the core of focusing tasks. Each specification consists of *input*, *output*, and a (functional) *criterion* that relates input to output (see definition 3.2.1). Whereas input and output are common for specifications, the focusing criterion is an additional aspect which is essential to precisely define what focusing solutions have to accomplish. The focusing criterion determines the overall purpose of a focusing task and its solution.

Definition 3.2.1 (Focusing Specification)

A focusing specification is a tuple $(f_{in}, f_{out}, f(f_{in}, f_{out}))$ with a focusing input f_{in} (see definition 3.2.2, page 48), a focusing output f_{out} (see definition 3.2.3, page 50), and a focusing criterion $f(f_{in}, f_{out})$ (see definition 3.2.4, page 50).

Each component of focusing specifications takes on one of several possible values. Combinations of single values for each component form concrete focusing specifications. Figure 3.2 outlines possible values for the components of focusing specifications. These components and their values span a cube of 18 concrete focusing specifications. At the beginning of KDD projects, the whole space of focusing specifications is possible. Step by step, the data mining engineer decides the necessity of focusing and restricts available focusing specifications by analyzing the focusing context. If this process defines a specific value for a single component, the set of currently possible specifications becomes a subset of available specifications in the previous step. In terms of our illustration, the refinement process limits the set of possible focusing specifications from the entire cube to a single side in the first step, from a single side to an edge in the second step, and finally from an edge to a single point.

If we combine each component value, the refinement process results in a concrete focusing specification. For example, *focusing tuples on a constructed set according to representativeness*, or *focusing attributes on a subset according to relevance* correspond to textual descriptions of focusing specifications. Note, in complex KDD projects it is possible to define more than a single focusing task. Then, data preparation refers to several points in the cube of focusing tasks, and solutions for each of these tasks contribute to the overall challenge of data reduction in such projects.

3.2.1   Focusing Input

The original *focusing input* is usually a single database table, a pre-selected subset of a database table, or result of joins on different database tables. Thus, the overall input for an implementation of a particular focusing solution corresponds to a (database) table. Each table is composed of a set of attributes and a set of tuples, and each attribute has its own domain of possible values. These three components of tables result in three different values for the focusing input of focusing specifications.

The focusing input is either a set of attributes, a set of tuples, or a set of values of a specific attribute domain (see definition 3.2.2). Hence, the focusing input specifies which component of a (database) table is manipulated by a focusing solution in order to solve a concrete focusing task. The specific focusing input defines the main concern of a focusing task, and each focusing input results in different focusing tasks which need different focusing solutions.

Definition 3.2.2 (Focusing Input)

Assume a (database) table (A, T). A focusing input f_{in} is either a set of tuples $(f_{in} \subseteq T)$, a set of attributes $(f_{in} \subseteq A)$, or a set of values $(f_{in} \subseteq dom(a_j)$, $a_j \in A)$.

In case of focusing tuples, the focusing task aims at reducing the number of rows in a table. For this reason, we also call focusing tasks with a set of tuples as its input *focusing in depth*. In contrast, focusing attributes attempts to restrict the original number of columns in a table. For this input, we refer to the respective set of focusing tasks as *focusing in breadth*. Finally, the input component defines additional focusing tasks, if the input is a subset of possible values for an attribute. This focusing input neither leads to focusing in depth nor to focusing in breadth directly. Nevertheless, focusing the number of values in tables possibly results in both focusing in depth and focusing in breadth.

Note, values of focusing input do not exclude each other. For example, focusing on one input often indirectly leads to focusing on another input. The focusing input only specifies the primary purpose, not the overall focusing result. Moreover, in some KDD projects it is necessary to focus the original data set for more than one input. For example, if a (database) table contains both a huge number of tuples and a large number of attributes, the project probably requires focusing tuples and focusing attributes. Then, we define two separate focusing tasks for this project, although both tasks depend on each other and interact.

3.2.2 Focusing Output

The focusing input is the first restriction of the *focusing output*. If the focusing input is a set of tuples, the focusing output is also a set of tuples. Similarly, if the focusing input is a set of attributes or a set of values, the focusing output is a set of attributes or a set of values, respectively. Note, the overall output of implementations of focusing solutions still represents all components of (database) tables.

We further distinguish between the following three types of focusing outputs. The basic type of focusing outputs is a *simple subset* of the focusing input. In case of simple subsets, we do not define any additional constraints on the focusing output except that it is a subset of the input. The second type of outputs is a *constrained subset*. Whereas the first type allows any arbitrary subset of the focusing input as the focusing output, the second type restricts the focusing output to specific subsets that meet desired constraints. Hence, this second type is a specialization of the first type. Finally, the focusing output is a *constructed set*, if we allow the focusing output to include artificial representatives of the focusing input. This type of focusing output no longer corresponds to subsets of the focusing input but constructions of new entities.

Definition 3.2.3 summarizes these three different types of focusing outputs. Note, in case of constrained subsets and constructed sets, we assume predicates P or functions P. A predicate P defines the set of constraints which the focusing output meets. A function P specifies construction principles how to generate the focusing output from the focusing input. We discuss specific examples of predicates and functions in chapter 4.

Definition 3.2.3 (Focusing Output)

Assume a (database) table (A, T), and a focusing input f_{in}. A focusing output f_{out} is either a simple subset of the focusing input ($f_{out} \subseteq f_{in}$), or a constrained subset of the focusing input ($f_{out} \subset f_{in}$, and $P(f_{in}, f_{out})$), or a constructed set of artificial entities ($f_{out} \not\subset f_{in}$, and $P(f_{in}, f_{out})$). P is a predicate or a function that relates f_{in} to f_{out}.

3.2.3   Focusing Criterion

The third component of focusing specifications determines the relation between focusing input and focusing output. The *focusing criterion* directs the search from focusing input to focusing output. This criterion is essential since it defines the purpose of focusing and describes what focusing solutions have to accomplish. At an abstract level, we distinguish between two different criteria, *relevance* and *representativeness* (see definition 3.2.4). In general, relevance means that focusing restricts the available information in the focusing input to a subset which is still appropriate to achieve the data mining goal. In contrast, representativeness ensures that the focusing output still represents the entire information in the focusing input.

Definition 3.2.4 (Focusing Criterion)

Assume a (database) table (A, T), a focusing input f_{in}, and a focusing output f_{out}. A focusing criterion $f(f_{in}, f_{out})$ is either relevance *or* representativeness.

Specific focusing criteria are either predicates or functions. In case of predicates, focusing criteria evaluate to true, if the relation between focusing input and focusing output meets the criterion. In order to get more fine-grained evaluations, we also consider focusing criteria as functions which return positive real values. These functions enable comparisons of different pairs of focusing inputs and focusing outputs. At this level, more precise definitions of relevance and representativeness are not possible, since focusing criteria have to take into account focusing input, focusing output, and focusing context.

For example, focusing tuples and focusing attributes are completely different tasks. The definition of focusing criteria is consequently completely different as well. For instance, John *et al.* (1994) present definitions of relevance in case of focusing attributes. Example 3.2.1 illustrates the definition of a particular simple example which additionally assumes classification tasks. Here, an attribute is relevant, if at least one value in its attribute domain and at least one class label exist such that the probability of this value is positive and the probability of the class label given this value is different from the probability of the class label.

This means, an attribute is relevant, if at least one of its values contributes information to at least one class label.

Example 3.2.1 (Attribute Relevance)

Assume a (database) table (A, T), a class attribute a_N, an attribute $a_j \in A$, and a probability function p on $dom(a_j)$ and $dom(a_N)$.

a_j relevant $:\Longleftrightarrow$

$\exists a_{jk} \in dom(a_j) \; \exists a_{Nk'} \in dom(a_N)$:

$p(t_{ij} \equiv a_{jk}) > 0 \; \wedge \; p(t_{iN} \equiv a_{Nk'} \mid t_{ij} \equiv a_{jk}) \neq p(t_{iN} \equiv a_{Nk'})$

This simple example already illustrates that specific definitions of focusing criteria have to take into account focusing input, focusing output, and focusing context. Specific focusing criteria map the intuitive notion of relevance or representativeness to focusing input, focusing output, and focusing context as well as their concrete data structures.

Since the focusing context contains data mining goal, data characteristics, and data mining algorithm, each of these three components influences the definition of specific focusing criteria as well. For example, assume we consider a segmentation goal in credit scoring to identify similar profiles of good customers, in order to retain existing customers or to approach new customers which share one of these specific profiles. Then, only good customers are of interest rather than good and bad customers, and the definition of relevance in this context requires value *good* for attribute *credit* of a customer (see example 2.4.1, page 27).

Similarly, data characteristics influence the definition of relevance or representativeness. For example, assume we consider tables with high redundancy in terms of duplicate tuples, and the data mining goal is to identify all unique entries, in order to analyze variations among customer profiles. Then, focusing outputs are already representative, if they contain at least one tuple for each subset of redundant tuples.

In a similar way, the data mining algorithm, which we want to apply to achieve the data mining goal, is relevant for definitions of specific focusing criteria. For example, if we compare top down induction of decision trees and nearest neighbor classifiers, we observe that relative frequencies of attribute values are important for the induction of decision trees, whereas basic nearest neighbor classifiers do not care about frequencies but need at least a single tuple from each subset of sufficiently similar tuples which belong to the same class in order to gain maximum accuracy.

We discuss further aspects of data characteristics and data mining algorithms in relation to focusing tasks in the following section, and we return to the definition of focusing criteria at the end of this chapter.

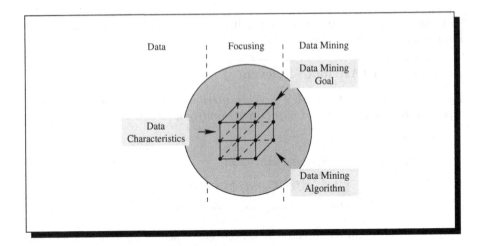

Figure 3.3: Focusing Context

3.3   Focusing Context

In the previous section, we defined focusing specifications as the first component of focusing tasks. Although this specification is the core of focusing tasks,
additional aspects influence focusing tasks and their solutions. Moreover, these
aspects play an important role for the definition of focusing specifications. For
example, if the original data set only contains a few tuples but many attributes,
the focusing input is a set of attributes rather than a set of tuples. Similarly,
if we plan to apply data mining algorithms which are not able to handle quantitative attributes, the focusing input is either a set of attributes and we have
to focus on qualitative attributes, or a set of values and we focus by discretization. In the second case, we get more than a single focusing specification, if the
original data set contains more than a single quantitative attribute.

Definition 3.3.1 (Focusing Context)

*Assume a (database) table (A, T). A focusing context is a tuple $(G, C(A, T), D)$
with G a data mining goal, $C(A, T)$ a set of data characteristics, and D a data
mining algorithm.*

As these examples illustrate, focusing depends on KDD phases before and
after data preparation, and the results of other KDD phases influence focusing
tasks. Here, we concentrate the *focusing context* on the most important aspects,
although all phases in KDD projects possibly affect focusing tasks as well as focusing solutions. Figure 3.3 outlines the most important aspects of the focusing
context (see definition 3.3.1). The focusing context mainly consists of data mining goal, data characteristics, and data mining algorithm. Specific instantiations

of these aspects summarize assumptions in specific KDD processes. For example, the context restricts the specific type of model which we want to generate. All in all, the focusing context guides the focusing process and determines its main purpose.

Within the focusing context, we also observe dependencies. Usually, we first analyze data characteristics in relation to the data mining goal. Then, we decide if we are able to achieve the data mining goal on the data. If we conclude that the data characteristics allow a solution of the data mining goal, we select appropriate data mining algorithms. As soon as we have selected data mining algorithms, we check if focusing is necessary. For instance, if we choose data mining algorithms which are not able to handle more than 1000 tuples but the data contains 2000 tuples, we define a focusing task with focusing input equals a set of tuples.

Note, whereas the data mining engineer defines focusing specifications during KDD projects, the focusing context is fixed before data preparation starts. The business analyst specifies data mining goals as translations of original business questions in the business understanding phase. After identification and selection of data sources, existing data determines the data characteristics. As soon as the data mining goal is clear and an initial data understanding phase helped to identify the data characteristics, the data mining engineer selects appropriate data mining algorithms for application. Hence, each component of the focusing context is fixed and no longer subject to modifications before the first data preparation phase starts. The more precise the context is, the better is the chance to select an appropriate focusing solution.

Once the focusing context is stable, we are able to define focusing criteria which specify requirements for appropriate focusing solutions. For example, if we deal with classification goals, and we are interested in highly accurate classifiers independent of the complexity of the data mining result, a focusing solution is only successful, if the application of data mining algorithms on the focusing output still generates sufficiently accurate classifiers.

Whereas we described each component of focusing contexts separately in the previous chapter, we now stress different roles of each component within the context of focusing. Although dependencies among components of focusing contexts are important as well, we emphasize on the relation between the focusing context and focusing tasks and their solutions. Since we fixed the data mining goal to classification at the end of chapter 2, we only consider data characteristics and different data mining algorithms as aspects of the focusing context. Although we do not discuss the influence of data mining goals on focusing, we notice that different data mining goals also result in different focusing contexts and specify different focusing tasks which require different focusing solutions.

In order to illustrate the relation between focusing context and focusing tasks as well as their solutions, we use the following example shown in figure 3.4.

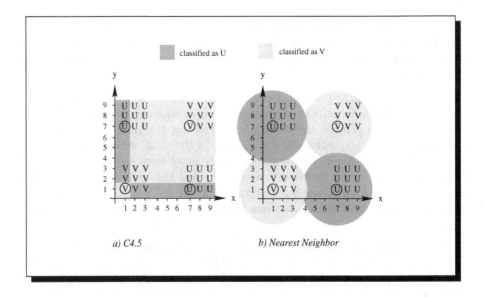

Figure 3.4: Focusing Context Example

Example 3.3.1 (Focusing Task)

Assume a (database) table (A, T) with $A = \{x, y, z\}$, $dom(x) = dom(y) = \{1, 2, 3, 7, 8, 9\}$, $dom(z) = \{U, V\}$, and $T = \{(1, 1, V), (1, 2, V), (1, 3, V), (1, 7, U), (1, 8, U), (1, 9, U), ...\}$ (see figure 3.4). Further assume the data mining goal G classification, and two data mining algorithms D_1 and D_2, C4.5 and NN.

We consider the focusing specification with $f_{in} = T$, i.e. a set of tuples, $f_{out} \subseteq T$, i.e., a simple subset, and a focusing criterion $f(f_{in}, f_{out})$ that defines representativeness in terms of the requirement to gain maximum classification accuracy, if we apply a data mining algorithm to the focusing output.

3.3.1 Data Characteristics

Data characteristics cover all types of information about the focusing input. For example, this information contains number of tuples, number of attributes, attribute types, statistical meta-data such as modus, mean, variance, and distributions, as well as quality of data mining results. Example 3.3.2 outlines some of these data characteristics for the focusing context in example 3.3.1.

Example 3.3.2 (Data Characteristics as Focusing Context)

Assume the focusing context in example 3.3.1. Then, the data characteristics contain 36 tuples, two quantitative predictive attributes and a qualitative class label with labels U and V, for instance.

In order to illustrate the influence of data characteristics on the definition of focusing tasks, we also need to consider specific implementations of data mining algorithms which we want to apply to achieve the data mining goal.

3.3.2 Data Mining Algorithms

We continue the discussion of the focusing task in example 3.3.1. Hence, we assume that we apply C4.5 as an implementation of top down induction of decision trees and algorithm NN as an implementation of nearest neighbor classifiers. Both implementations use different strategies to classify unseen tuples. Whereas C4.5 builds a decision tree using the focusing output as the training set, NN uses the focusing output to retrieve the most similar tuple for each tuple in the test set and returns the class label of this nearest neighbor as the predicted class of an unseen tuple.

Example 3.3.3 illustrates that the same focusing output is less appropriate in one focusing context but more appropriate in another focusing context. In this case, the specific strategies of classification in top down induction of decision trees and nearest neighbor classifiers result in completely different evaluations of the same focusing solution.

Example 3.3.3 (Data Mining Algorithms as Focusing Context)

Assume the focusing task in example 3.3.1, a focusing output $f_{out} = \{(1, 1, V),$ $(1, 7, U)$, $(7, 1, U)$, $(7, 7, V)\}$ (see circled tuples in figure 3.4), and a test set $T_{test} = T$.

Then, we observe that the same focusing output is less appropriate for C4.5 than for nearest neighbor according to the specified focusing criterion in example 3.3.1. Whereas C4.5 classifies 11 tuples in the test set as U (see dark gray area in figure 3.4) and the remaining tuples as V (see light gray area in figure 3.4) and yields a classification accuracy of 55%, the application of NN to the same focusing output classifies all tuples correctly and results in a classification accuracy of 100%.

In summary, it is important to consider all aspects of the focusing context which influence the definition of focusing tasks before the development or selection of appropriate focusing solutions.

3.4 Focusing Success

Focusing success of focusing solutions depends on the quality of the focusing output. This quality again depends on focusing input, focusing criterion, and focusing context. In this section, we specify *evaluation criteria* to measure the quality of focusing outputs. These criteria enable comparative studies of different focusing solutions.

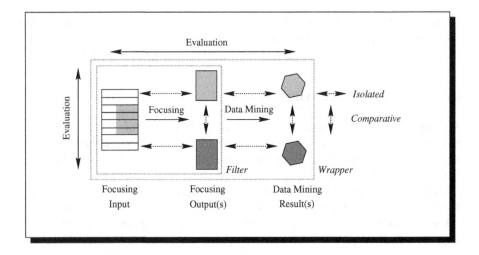

Figure 3.5: Evaluation Strategies for Focusing Success

Evaluation Strategies

The potential usage of evaluation criteria is two-fold. We either measure the quality of focusing outputs at the end of applications of focusing solutions, or evaluation is integrated into a focusing solution and we reject intermediate focusing outputs and generate new focusing outputs, if the quality of the focusing output is not sufficient yet. For our purposes, we focus on the first type of evaluation.

The naive approach to evaluation presents focusing outputs or data mining results to domain experts. These experts measure the quality of focusing solutions according to their own preferences. For example, classification rules generated on focusing outputs possibly are more interesting than rules produced on the entire focusing input. However, this approach is not feasible in general cases. Human expert's judgments are often subject to personal preferences and lead to less objective comparisons of focusing solutions. We define evaluation criteria that automatically measure focusing success independent of human expert's opinions.

Figure 3.5 presents an overview of different evaluation strategies. Evaluation criteria that realize notions of relevance or representativeness mainly differ along the following three aspects:

- *Qualitative and Quantitative Evaluation*

  Qualitative evaluation states whether focusing outputs are *good* or *bad*. Instead, quantitative evaluation expresses the quality of focusing solutions in terms of numeric values. Consequently, quantitative evaluation allows more fine-grained comparisons of focusing solutions, since rankings of the quality of focusing outputs are possible.

- *Filter and Wrapper Evaluation*

  Filter evaluation only considers focusing outputs without any relation to data mining and its results. In contrast, wrapper evaluation takes into account the data mining phase and its results (John *et al.*, 1994; Kohavi, 1995a).[1] Filter evaluation emphasizes statistical aspects and data characteristics, whereas wrapper evaluation specifically regards the data mining phase and compares data mining results on focusing outputs in order to measure the quality of focusing solutions. Filter evaluation is generally more efficient and less time-consuming than wrapper evaluation since we do not necessarily need to apply data mining algorithms. However, wrapper evaluation is more appropriate for comparisons of focusing solutions for data mining.

- *Isolated and Comparative Evaluation*

  Isolated evaluation measures focusing success of single focusing solutions. In this case, we either concentrate on the focusing output and the data mining result, or we compare a single focusing output and its data mining result to the focusing input and the data mining result on the focusing input. Comparative evaluation relates two or more focusing outputs to each other. This strategy allows to measure the quality of focusing solutions relatively in comparison to other focusing solutions.

Since we are interested in rankings of different focusing solutions, we prefer the development of quantitative evaluation criteria. In the following, we discuss filter and wrapper evaluation as well as different aspects of both. We also consider isolated as well as comparative evaluation in both cases. Thus, we define quantitative filter and wrapper evaluation criteria in isolated and comparative modes. At the end, we also specify combined evaluation criteria which cumulate more than a single evaluation strategy. Note, we only consider the selected focusing tasks. We do not discuss evaluation criteria for different focusing contexts.[2]

3.4.1 Filter Evaluation

Filter evaluation compares statistical data characteristics of focusing input and focusing output, or statistical characteristics of different focusing outputs in relation to the focusing input. Each focusing input comprises specific characteristics, and we consider the resulting meta-data to measure representativeness of focusing outputs. A natural evaluation criterion for focusing success requires

---

[1]Note, the original definition of wrappers includes iterative improvement of focusing outputs.

[2]For example, see Michie *et al.* (1994) for an evaluation of statistical and machine learning techniques in different domains.

that focusing outputs still represent the same or similar meta-data as the focusing input. In some cases, this approach also allows small deviations within
pre-specified limits.

We distinguish three types of statistical characteristics for filter evaluation.
The first type of characteristics describes characteristic values such as *modus*,
mean, and *variance* of attribute values. Whereas we apply the modus in case
of qualitative attributes, we use mean and variance in case of quantitative attributes. The second type considers the *distribution* of attribute values for single
attributes. We measure focusing success according to distributions by considering the distribution of attribute values for each attribute within the focusing
output separately. Finally, we also take into account *joint distributions* of attribute values for more than a single attribute.

Filter evaluation uses these three classes of statistical characteristics to measure focusing success in the following way:

- *Modus, Mean, and Variance*

  The modus of an attribute is the most often occurring value in its domain.
  We expect that focusing solutions with high quality select focusing outputs
  which represent the same modus as the focusing input.

  The mean of a quantitative attribute is the average value of all occurring
  values in its domain. In this case, we expect that the average value in
  focusing outputs is approximately the same as in the focusing input for
  focusing solutions with high quality.

  For quantitative attributes, we also consider variances among occurring
  values. This allows to measure the variation from average values within
  sets of occurring values. Focusing outputs of focusing solutions with high
  quality represent approximately the same variance as the focusing input.

- *Distributions*

  Whereas characteristic values consider only aggregated information of occurring values in (database) tables, distributions measure frequencies of
  occurring values for each value separately. The quality of focusing solutions is high, if the respective focusing output represents approximately
  the same distribution as the focusing input.

- *Joint Distributions*

  Similar to distributions, joint distributions take into account frequencies of
  occurring attribute values. Rather than considering single attributes and
  their domains, joint distributions inspect all attributes in a (database) table simultaneously. This case is similar to the previous one since we analyze
  combinations of values for different attributes and take these combinations
  as single values. Focusing solutions of high quality represent approximately
  the same joint distributions as the focusing input.

In the following subsections, we describe hypothesis testing in statistics to analyze characteristic values and their representations in focusing outputs and define specific tests for each of the three aspects of statistical characteristics in filter evaluation. We use hypothesis tests to define specific filter evaluation criteria which measure focusing success in terms of statistical representativeness.

3.4.1.1 Hypothesis Testing

The key procedure in statistics to analyze assumptions about focusing outputs (samples) and their relation to focusing inputs (populations) is *hypothesis testing*. Since we use hypothesis testing to define filter evaluation criteria that take into account statistical characteristics, we briefly describe principles of hypothesis testing in statistics. The overall goal of hypothesis testing is to decide which of two complementary hypotheses is true. In general, *hypotheses* are statements about population parameters. In our context, the focusing input represents the population and parameters refer to statistical characteristics of the focusing input.[3]

Definition 3.4.1 (Statistical Hypotheses)

Assume a (database) table (A, T), *a set of data characteristics* $C(A, T)$, *a parameter* $\theta \in C(A, T)$, $\Theta_0 \subset dom(\theta)$, *and* $\Theta_0^c = dom(\theta) - \Theta_0$.

$H_0 : \theta \in \Theta_0$ *is the* null hypothesis.

$H_1 : \theta \in \Theta_0^c$ *is the* alternative hypothesis.

Assume a (database) table (A, T), *a set of data characteristics* $C(A, T)$, *a parameter* $\theta \in C(A, T)$, *and* $\Theta_0 \in dom(\theta)$.

$H_0 : \theta \geq \Theta_0,$ $H_1 : \theta < \Theta_0$ *are one-sided hypotheses.*

$H_0 : \theta = \Theta_0,$ $H_1 : \theta \neq \Theta_0$ *are two-sided hypotheses.*

Definition 3.4.1 specifies characteristics of statistical hypotheses. The first part in this definition is more general than the second, since the definition of null and alternative hypotheses considers any concrete subset of the domain of a population parameter. In contrast, the second definition of null and alternative hypotheses uses a specific value in the domain of a population parameter and either formulates a test which only compares the sample parameter to the specific value as a threshold (*one-sided* hypotheses) or requires an exact coincidence between sample and population values for this parameter (*two-sided* hypotheses). Since we attempt to represent information in the focusing input as good as possible, we focus our attention on two-sided hypotheses.

---

[3] As we already pointed out, the focusing input itself is only a sample of a larger population.

Table 3.1: Two Types of Errors in Hypothesis Testing

	Accept H_0	Reject H_0
H_0 true	correct	type I error (α-error)
H_1 true	type II error (β-error)	correct

Hypothesis testing procedures or *hypothesis tests* are rules that specify in which situations to accept H_0 and to reject H_1 or to reject H_0 and to accept H_1. Typically, hypothesis tests are specified in terms of a *test statistic* which is a function of focusing input and focusing output and a *rejection region* which defines which values of the test statistic lead to rejections of the null hypothesis. The crucial part in hypothesis testing is consequently the definition of an appropriate test statistic and a suitable rejection region.

Hypothesis tests eventually make one of two types of errors. If the null hypothesis is true but a hypothesis test incorrectly rejects this hypothesis, then this test makes a *type I error* (α-error). On the other hand, if the alternative hypothesis is true but a hypothesis test incorrectly accepts the null hypothesis, this test makes a *type II error* (β-error). Table 3.1 summarizes correct test decisions and both types of errors.

In hypothesis testing, we are interested in tests that are unlikely to make one of these two types of errors. If the null hypothesis is true, we call the probability α that the test statistic value is in the rejection region and that we incorrectly conclude to reject the null hypothesis *significance level* of the test. The significance level of a test corresponds to the probability of a type I error in table 3.1. In the literature, hypothesis testing also uses the term *confidence level*. The confidence level is $1 - \alpha$ and refers to the chance of failing to reject the null hypothesis when it is true.

Definition 3.4.2 summarizes the procedure of hypothesis testing. If we consider a null hypothesis versus an alternative hypothesis, we specify a test statistic for the null hypothesis which is a function of focusing input and focusing output and an *expected distribution* of this test statistic. In order to decide whether to accept or reject the null hypothesis, we compare the absolute value of the test statistic with the expected distribution. If the absolute value of the test statistic is larger than the expected distribution, we reject the null hypothesis at confidence level $1 - \alpha$. Hence, the rejection region of a hypothesis test is the set of test statistic values which are larger than the respective expected distribution.

Definition 3.4.2 (Hypothesis Testing)

Assume a (database) table (A, T), a focusing input $f_{in} \subseteq T$, a focusing output $f_{out} \subseteq f_{in}$, a null hypothesis H_0, an alternative hypothesis H_1, a test statistic $s : T^{\subseteq} \times T^{\subseteq} \longrightarrow] - \infty; \infty[$ for H_0, and an expected distribution $d :]0; 1[\times T^{\subseteq} \longrightarrow] - \infty; \infty[$ for s at confidence level $1 - \alpha$. We define the rejection region for H_0 as

$$\{s(f_{in}, f_{out}) \mid \mid s(f_{in}, f_{out}) \mid > d(\alpha, f_{in})\}, \text{ or}$$

$$\{s(f_{in}, f_{out}) \mid \mid s(f_{in}, f_{out}) \mid > d(\alpha, f_{out})\}, \text{ respectively,}$$

i.e., we reject H_0 at confidence level $1-\alpha$, if and only if $\mid s(f_{in}, f_{out}) \mid > d(\alpha, f_{in})$ or $\mid s(f_{in}, f_{out}) \mid > d(\alpha, f_{out})$, respectively.

We utilize statistical hypothesis testing for qualitative comparisons between focusing input and focusing output. If we also want to estimate the amount of deviation between characteristics in the focusing input and their representation in focusing outputs, we compute the fraction of test statistic and expected distribution. If this fraction is larger than 1, we reject the null hypothesis. The higher the fraction value is, the larger is the deviation between a characteristic value in the focusing input and its representation in the focusing output. We apply qualitative and quantitative aspects of hypothesis testing for definitions of filter evaluation criteria.

Note, in statistics there are two different types of hypothesis tests. The first type compares two different focusing outputs from the same focusing input and decides whether both focusing outputs represent the same population. The second type relates single focusing outputs to the focusing input and tests whether this focusing output appropriately represents the characteristics in the focusing input. We focus our attention on the second type of tests, since we are more interested in comparisons between focusing input and focusing output and the level of representativeness of focusing outputs.

3.4.1.2 Hypothesis Testing for Meta-Data

In the following, we define specific hypothesis tests in terms of test statistics and expected distributions for means, variances, and distributions.[4] Since we assume that the reader is familiar with statistics, we only cite these hypothesis tests without detailed explanations (see Mason *et al.* (1989), for example).[5]

Note, in order to apply this hypothesis test for variances, we assume normal distribution of population and samples. According to the *central limit property*, this assumption is valid, if tuples in the sample are independent and from the same probability distribution, and the sample size is sufficiently large.

---

[4]Note, there is no hypothesis test for the modus of attributes.

[5]Note, the design of hypothesis testing in statistics usually requires random samples. In theory, statistics does not completely justify results of applications of hypothesis tests to focusing outputs from deterministic focusing solutions.

Definition 3.4.3 (Hypothesis Test for Means)

Assume a (database) table (A, T), a focusing input $f_{in} \subseteq T$, a focusing output $f_{out} \subseteq f_{in}$ with $m := | f_{out} |$, and a quantitative attribute $a_j \in A$. We define the null hypothesis and alternative hypothesis for means as

$$H_0 : \mu_j(f_{in}) = \mu_j(f_{out}), \quad \text{and} \quad H_1 : \mu_j(f_{in}) \neq \mu_j(f_{out}).$$

We define the test statistic and the expected distribution for means as two functions

$$s_{S_j} : T^{\subseteq} \times T^{\subseteq} \longrightarrow] - \infty; \infty[,$$

$$s_{S_j}(f_{in}, f_{out}) := \frac{\mu_j(f_{out}) - \mu_j(f_{in})}{\sigma_j(f_{out})} \cdot \sqrt{m},$$

$$d_{S_j} :]0; 1[\times T^{\subseteq} \longrightarrow] - \infty; \infty[,$$

$$d_{S_j}(\alpha, f_{out}) := \begin{cases} z_{1-\frac{\alpha}{2}}, & \text{if } m \geq 30 \\ t_{1-\frac{\alpha}{2}}(m - 1), & \text{else} \end{cases}$$

$z_{1-\frac{\alpha}{2}}$ *is the* $(1-\frac{\alpha}{2})$*-quantile of the standard normal distribution, and* $t_{1-\frac{\alpha}{2}}(m-1)$ *is the* $(1 - \frac{\alpha}{2})$*-quantile of the student's t-distribution with* $m - 1$ *degrees of freedom. In order to simplify some of the following definitions, we also define a function*

$$q_{S_j} :]0; 1[\times T^{\subseteq} \times T^{\subseteq} \longrightarrow] - \infty; \infty[,$$

$$q_{S_j}(\alpha, f_{in}, f_{out}) := \frac{s_{S_j}(f_{in}, f_{out})}{d_{S_j}(\alpha, f_{out})}.$$

Definition 3.4.4 (Hypothesis Test for Variances)

Assume a (database) table (A, T), a focusing input $f_{in} \subseteq T$, a focusing output $f_{out} \subseteq f_{in}$ with $m := | f_{out} |$, and a quantitative attribute $a_j \in A$. We define the null hypothesis and alternative hypothesis for variances as

$$H_0 : \sigma_j(f_{in})^2 = \sigma_j(f_{out})^2, \quad \text{and} \quad H_1 : \sigma_j(f_{in})^2 \neq \sigma_j(f_{out})^2.$$

We define the test statistic and the expected distribution for variances as two functions

$$s_{V_j} : T^{\subseteq} \times T^{\subseteq} \longrightarrow] - \infty; \infty[, \qquad s_{V_j}(f_{in}, f_{out}) := \frac{(m - 1) \cdot \sigma_j(f_{out})^2}{\sigma_j(f_{in})^2},$$

$$d_{V_j} :]0; 1[\times T^{\subseteq} \longrightarrow] - \infty; \infty[, \quad d_{V_j}(\alpha, f_{out}) := \chi^2_{1-\frac{\alpha}{2}}(m - 1).$$

$\chi^2_{1-\frac{\alpha}{2}}(m - 1)$ *is the* $(1 - \frac{\alpha}{2})$*-quantile of the* χ^2*-distribution with* $m - 1$ *degrees of freedom. In order to simplify some of the following definitions, we also define a function*

$$q_{V_j} :]0; 1[\times T^{\subseteq} \times T^{\subseteq} \longrightarrow] - \infty; \infty[,$$

$$q_{V_j}(\alpha, f_{in}, f_{out}) := \frac{s_{V_j}(f_{in}, f_{out})}{d_{V_j}(\alpha, f_{out})}.$$

Note also, although the definition only contains the $(1 - \frac{\alpha}{2})$-quantile of the χ^2-distribution with $m - 1$ degrees of freedom, we also have to compare the absolute value of the test statistic and the $\frac{\alpha}{2}$-quantile of the χ^2-distribution with $m - 1$ degrees of freedom to perform a complete hypothesis test for variances.

Definition 3.4.5 (Hypothesis Test for Distributions)

Assume a (database) table (A, T), a focusing input $f_{in} \subseteq T$ with $M := | f_{in} |$, a focusing output $f_{out} \subseteq f_{in}$ with $m := | f_{out} |$, and an attribute $a_j \in A$. We define the null hypothesis and alternative hypothesis for distributions as

$$H_0 : \quad \forall a_{jk} \in dom(a_j) : \ n_{jk}(f_{in}) = n_{jk}(f_{out}), \text{ and}$$

$$H_1 : \quad \exists a_{jk} \in dom(a_j) : \ n_{jk}(f_{in}) \neq n_{jk}(f_{out}).$$

We define the test statistic and the expected distribution for distributions as two functions

$$s_{D_j} : T^{\subseteq} \times T^{\subseteq} \longrightarrow] - \infty; \infty[,$$

$$s_{D_j}(f_{in}, f_{out}) := \sum_{k=1}^{N_j} \frac{\left(n_{jk}(f_{out}) - m \cdot \dfrac{n_{jk}(f_{in})}{M} \right)^2}{m \cdot \dfrac{n_{jk}(f_{in})}{M}},$$

$$d_{D_j} :]0; 1[\times T^{\subseteq} \longrightarrow] - \infty; \infty[, \quad d_{D_j}(\alpha, f_{in}) := \chi^2_{1-\alpha}(N_j - 1).$$

$\chi^2_{1-\alpha}(N_j - 1)$ *is the $(1 - \alpha)$-quantile of the χ^2-distribution with $N_j - 1$ degrees of freedom. In order to simplify some of the following definitions, we also define a function*

$$q_{D_j} :]0; 1[\times T^{\subseteq} \times T^{\subseteq} \longrightarrow] - \infty; \infty[,$$

$$q_{D_j}(\alpha, f_{in}, f_{out}) := \frac{s_{D_j}(f_{in}, f_{out})}{d_{D_j}(\alpha, f_{out})}.$$

Note, if we use this hypothesis test for distributions, we have to ensure that $m \cdot n_{jk}(f_{in}) \geq 5$ for all values $a_{jk} \in dom(a_j)$. For quantitative attributes, we propose to utilize discretization approaches (see section 4.2.2.5, page 128ff) and unification of intervals, if the focusing input violates this condition. For qualitative attributes, we suggest to group values, if the expected frequency is less than 5.

Definition 3.4.6 (Hypothesis Test for Joint Distributions)

Assume a (database) table (A, T), a focusing input $f_{in} \subseteq T$ with $M := \mid f_{in} \mid$, and a focusing output $f_{out} \subseteq f_{in}$ with $m := \mid f_{out} \mid$. We define the null hypothesis and alternative hypothesis for joint distributions as

$$H_0 : \quad \forall a_{1k_1} \in dom(a_1), \dots, \forall a_{jk_j} \in dom(a_j), \dots, \forall a_{Nk_N} \in dom(a_N) :$$

$$n_{k_1 \dots k_j \dots k_N}(f_{in}) = n_{k_1 \dots k_j \dots k_N}(f_{out}), \text{ and}$$

$$H_1 : \quad \exists a_{1k_1} \in dom(a_1), \dots, \exists a_{jk_j} \in dom(a_j), \dots, \exists a_{Nk_N} \in dom(a_N) :$$

$$n_{k_1 \dots k_j \dots k_N}(f_{in}) \neq n_{k_1 \dots k_j \dots k_N}(f_{out}).$$

We define the test statistic and the expected distribution for joint distributions as two functions

$$s_J : T^{\subseteq} \times T^{\subseteq} \longrightarrow \,] - \infty; \, \infty[,$$

$$s_J(f_{in}, f_{out})$$

$$:= \sum_{k_1=1}^{N_1} \cdots \sum_{k_j=1}^{N_j} \cdots \sum_{k_N=1}^{N_N} \frac{\left(n_{k_1 \dots k_j \dots k_N}(f_{out}) - m \cdot \dfrac{n_{k_1 \dots k_j \dots k_N}(f_{in})}{M} \right)^2}{m \cdot \dfrac{n_{k_1 \dots k_j \dots k_N}(f_{in})}{M}},$$

$$d_J : \,]0; \, 1[\, \times T^{\subseteq} \longrightarrow \,] - \infty; \, \infty[, \quad d_J(\alpha, f_{in}) := \chi^2_{1-\alpha}(\prod_{j=1}^{N} N_j - 1).$$

$\chi^2_{1-\alpha}(\prod\limits_{j=1}^{N} N_j - 1)$ *is the $(1 - \alpha)$-quantile of the χ^2-distribution with $\prod\limits_{j=1}^{N} N_j - 1$ degrees of freedom. In order to simplify some of the following definitions, we also define a function*

$$q_J : \,]0; \, 1[\, \times T^{\subseteq} \times T^{\subseteq} \longrightarrow \,] - \infty; \, \infty[, \quad q_J(\alpha, f_{in}, f_{out}) := \frac{s_J(f_{in}, f_{out})}{d_J(\alpha, f_{out})}.$$

Note, if we use this hypothesis test for joint distributions, we have to ensure that $m \cdot n_{k_1 \dots k_j \dots k_N}(f_{in}) \geq 5$ for all combinations of values $a_{1k_1} \in dom(a_1), \dots,$ $a_{jk_j} \in dom(a_j), \dots, a_{Nk_N} \in dom(a_N)$. We propose to utilize the same strategies as for hypothesis tests for distributions of single attributes, if this assumption is not true in the focusing input.

3.4.2   Wrapper Evaluation

Wrapper evaluation measures the quality of focusing solutions indirectly by analyzing data mining results on focusing outputs. This type of evaluation of focusing outputs considers the quality of data mining results rather than statistical representativeness of data characteristics. The advantage is that we explicitly consider the original purpose of focusing solutions for data mining, since we apply focusing solutions in order to enable or to improve data mining in KDD. Focusing outputs of statistically high quality do not necessarily yield good data mining results. Vice versa, focusing outputs that yield good data mining results do not necessarily have high statistical quality.

In focusing contexts with huge focusing inputs, we are usually not able to compare data mining results on the entire focusing input and data mining results on focusing outputs. Instead, we only consider the quality of data mining results on single focusing outputs, or we compare data mining results on different focusing outputs. However, the experimental results chapter also contains domains where the focusing input size is small enough to apply data mining algorithms to the entire focusing input. This enables isolated comparisons of single focusing outputs and the entire focusing input.

Wrapper evaluation indirectly measures focusing success by analyses of data mining results. The quality of data mining results in classification mainly depends on the following aspects:

- *Execution Time and Storage Requirements*

  The more efficient applications of data mining algorithms on focusing outputs are, the higher is the quality of focusing solutions. Efficiency is usually measured by algorithmic complexity, execution time, and storage requirements. Since focusing solutions hardly alter algorithmic complexity of data mining algorithms, we only consider execution time and storage requirements. Since focusing solutions require extra costs, we not only take into account efficiency of data mining algorithms but also execution time and storage requirements of focusing solutions.

- *Accuracy*

  The more accurate resulting classifiers from applications of data mining algorithms to focusing outputs are, the higher is the quality of focusing solutions. Accuracy is usually measured by the relative number of correct classifications or the relative number of incorrect classifications, respectively.

- *Complexity*

  The less complex data mining results on focusing outputs are, the higher is the quality of focusing solutions. If data mining results are complex (e.g., in terms of the number of classification rules or the number of conditions in classification rules), the classifier application step in classification is more complex as well. It either takes more time to apply complex classifiers, or the classifier is no longer readable by any non-automatic mechanism.

If two focusing outputs are similar in one aspect, a focusing output still outperforms the other one, if its evaluation is superior on an additional aspect.

For example, if accuracies are the same, less complex data mining results are preferable. In other contexts, evaluation weighs all aspects to represent the importance of each aspect. For example, if accuracy is less important than complexity, comparative evaluation of focusing outputs which result in small decreases in accuracy for simple classifiers and focusing outputs which yield slightly higher accuracies but much more complex classifiers ranks the quality of the first focusing solutions higher than the second. In general terms, evaluation criteria define combined cost functions which are able to take into account any aspect of interest (Nakhaeizadeh & Schnabl, 1997).

Now, we define intermediate measures which allow to quantify each of the three aspects of data mining result quality for classification goals. We use these measures to define specific wrapper evaluation criteria which measure focusing success in terms of the quality of data mining results on focusing outputs.

3.4.2.1   Execution Time and Storage Requirements

Execution time measures running time of applications of focusing solutions to the focusing input and running time of applications of data mining algorithms to focusing input and focusing output, respectively. Since running time is the time which processes run on machines, and computers allow to measure the amount of time which processes need from their start to their termination, we are able to realize a function t in order to measure execution time (see definition 3.4.7).

Definition 3.4.7 (Execution Time)

Assume a (database) table (A, T), a focusing input $f_{in} \subseteq T$, a focusing solution $F : T^{\subseteq} \longrightarrow T^{\subseteq}$, a focusing output $f_{out} = F(f_{in}) \subseteq f_{in}$, and a data mining algorithm D. We define execution time as a function

$$t : \{F, D\} \times T^{\subseteq} \longrightarrow \,]0; \infty[$$

that measures running time of the focusing solution applied to the focusing input, and running time of data mining on the focusing input or on the focusing output, respectively.

In order to compare focusing solutions according to execution time, we take into account both, execution time for focusing and execution time for data mining. In isolated comparisons, we relate the sum of execution time for focusing on the focusing input and data mining on the focusing output to execution time for data mining on the focusing input. In comparative evaluations, we compare the sum of execution time for focusing on the focusing input and data mining on the focusing output for two or more different focusing solutions and their resulting focusing outputs.

Definition 3.4.8 (Storage Requirements)

Assume a (database) table (A, T), a focusing input $f_{in} \subseteq T$, a focusing solution $F : T^{\subseteq} \longrightarrow T^{\subseteq}$, a focusing output $f_{out} = F(f_{in}) \subseteq f_{in}$, and a data mining algorithm D. We define storage requirements as a function

$$m : \{F, D\} \times T^{\subseteq} \longrightarrow \;]0; \; \infty[,$$

$$m(F, f_{in}) := \mid f_{in} \mid, \quad m(D, f_{in}) := \mid f_{in} \mid, \quad m(D, f_{out}) := \mid f_{out} \mid$$

that measures storage requirements of the focusing solution applied to the focusing input, and storage requirements of data mining on the focusing input or on the focusing output, respectively.

Storage requirements of focusing solutions and data mining algorithms mainly depend on the number of attributes and the number of tuples in the focusing input and in the focusing output, respectively. Since focusing solutions for the selected focusing tasks do not reduce the number of attributes, we neglect the factor of storage requirements for attributes since this factor is the same for all focusing solutions and data mining algorithms. Instead, the number of tuples in focusing outputs is smaller than the number of tuples in the focusing input. Hence, the interesting difference in storage requirements for focusing solutions and data mining algorithms is the difference in the number of tuples. Consequently, we measure storage requirements by the number of tuples in the focusing input and in the focusing output, respectively (see definition 3.4.8).

3.4.2.2 Accuracy

Classification accuracy is usually measured according to predictive accuracies of classifiers. The predictive accuracy is the number of correct classifications, if we apply classifiers to sets of unseen tuples. Several mechanisms to estimate predictive accuracies exist (Kohavi, 1995b):

- *Train and Test*

  If we utilize train and test estimations for classification accuracies, we separate the original data set into two non-overlapping subsets, the training set and the test set. It is common to use about $\frac{2}{3}$ of the data as the training set and the remaining $\frac{1}{3}$ as the test set. Then, we apply data mining algorithms to the training set to generate classifiers, and use the resulting classifiers to assign class labels to tuples in the test set. The number of correct classifications in the test set yields an estimate for the classification accuracy of the classifier. Since the separation of the original data set into training and test set is usually done with random sampling, it is advisory to repeat the train and test mechanism several times and to compute the final estimation for classification accuracies by averaging the resulting accuracies from each run.

- *Cross-Validation*

  Cross-validation procedures to estimate classification accuracies are an extension of the train and test strategy. We randomly separate the original data set into L non-overlapping subsets. Then, we use the train and test approach L times. Each time, we select a different subset as the test set and the unification of all remaining $L - 1$ subsets as the training set. The resulting estimate for classification accuracies is the sum of estimations of classification accuracies from all L runs using the train and test method.

- *Bootstrapping*

  The bootstrap approach (Efron & Tibshirani, 1993) draws a random sample of size M with replacement, if the original data set contains M tuples. This sample constitutes the training set whereas the rest of the original data forms the test set. Then, we proceed the estimation of classification accuracies as in the train and test method. We repeat the bootstrap step b times for a pre-specified number b of bootstrap samples. The overall estimation of classification accuracies is the cumulation of train and test estimations from each of the b separations including a component that also measures accuracies on the training set rather than only on the test set.

We decide to use the first alternative and to measure classification accuracies with a train and test strategy. Although cross-validation and bootstrapping sometimes yield more reliable and less biased estimates, we prefer the train and test strategy for simplicity. If we use cross-validation or bootstrapping, we have to apply focusing solutions to each training set separately, run data mining algorithms on each focusing output, and test the resulting classifiers on the remaining test set. This procedure is quite complex and contains an additional bias since focusing solutions behave differently on each training set. Thus, we use the first alternative and take single training sets as the focusing input and test the resulting classifiers on single test sets. Since we are interested in performance of focusing solutions rather than accuracies of data mining algorithms, this strategy is appropriate. All focusing solutions suffer from the same bias, if the single train and test split is not representative for the classification goal.

Definition 3.4.9 (Accuracy and Error Rate)

Assume a (database) table (A, T), a classifier $\phi : T_{[1;\ N-1]} \longrightarrow dom(a_N)$, and a test set $T_{test} \subseteq T$. We define the accuracy of ϕ on T_{test} as a function

$$a : \phi \times T^{\subseteq} \longrightarrow [0;\ 1],$$

$$a(\phi, T_{test}) := \frac{\mid \{t_i \in T_{test} \mid \phi(t_{i[1;\ N-1]}) \equiv t_{iN}\} \mid}{\mid T_{test} \mid}.$$

Correspondingly, we define the error rate of ϕ on T_{test} as a function

$$e : \phi \times T^{\subseteq} \longrightarrow [0;\ 1], \quad e(\phi, T_{test}) := 1 - a(\phi, T_{test}).$$

In order to evaluate focusing solutions according to quality of data mining results in classification, we define an accuracy and a corresponding error rate function (see definition 3.4.9). Accuracy and error rate relate the number of correct classifications on the test set to its size. Hence, accuracy measures the relative number of correct classifications, whereas the error rate counts the relative number of incorrect classifications.

Note, this definition does not take into account classification costs. In some domains, costs of classification errors vary, and it is more appropriate to define weighted accuracy and error rate functions that take into account these costs (e.g., Provost & Fawcett, 1997). For example, in credit scoring misclassification costs in case of good customers who are classified as bad are usually lower than costs in case of classifying bad customers as good.

3.4.2.3 Complexity

The complexity of data mining results depends on the representation of resulting classifiers. Thus, we have to distinguish between different data mining algorithms. For example, in case of rule learning algorithms we measure complexity of rules, whereas in case of top down induction of decision trees we measure complexity of decision trees. For nearest neighbor classifiers, we need an additional measure of complexity, too.

Since we selected top down induction of decision trees and nearest neighbor classifiers as classification algorithms, we only consider complexity of data mining results for these two types of algorithms. For top down induction of decision trees, complexity of data mining results depends on the number of levels and the number of nodes in resulting decision trees (see definition 3.4.10). Both numbers influence understandability and applicability of data mining results. The number of levels corresponds to the number of attribute value tests for classification of unseen tuples in the worst case, when classification of unseen tuples tests all attribute values along the longest path in a decision tree. The number of nodes indicates the number of different attribute test outcomes. The number of outcomes does not influence the classifier application step but designates the complexity of decision trees in terms of their size. The more nodes decision trees have, the less comprehensive and consequently less understandable are the resulting classifiers.

Definition 3.4.10 (Decision Tree Complexity)

Assume a (database) table (A, T), and a decision tree $\phi : T_{[1;\ N-1]} \longrightarrow dom(a_N)$. We define the complexity of ϕ as two functions

$$l : \phi \longrightarrow]0;\ \infty[, \quad and \quad n : \phi \longrightarrow]0;\ \infty[$$

that count the number of levels and the number of nodes in the decision tree, respectively.

Table 3.2: Isolated and Comparative Filter Evaluation Criteria

	Modus and Mean	Variance	Distribution	Joint Distribution
isolated	$E_{S_j}^{\leftrightarrow}$	$E_{V_j}^{\leftrightarrow}$	$E_{D_j}^{\leftrightarrow}$	$E_{J}^{\leftrightarrow}$
comparative	$E_{S_j}^{\updownarrow}$	$E_{V_j}^{\updownarrow}$	$E_{D_j}^{\updownarrow}$	$E_{J}^{\updownarrow}$

In case of nearest neighbor classifiers, complexity of data mining results is inherent in the size of the training set and the complexity of the similarity measure. Since the complexity of similarity measures is independent of the number of tuples used for classification and only decreases in case of focusing on attributes, we ignore this component of nearest neighbor classifiers' complexity. Hence, we only consider the number of tuples in the training set which is again the same as the storage requirements (see definition 3.4.8, page 67).

3.4.3   Evaluation Criteria

This section summarizes evaluation aspects of the previous two sections and defines specific evaluation criteria. First, we describe evaluation criteria for each aspect of filter evaluation as well as wrapper evaluation in isolated and comparative modes. Step by step, we combine these measures to cumulated evaluation criteria which take into account more than single evaluation aspects. All definitions of evaluation criteria return smaller values for better focusing solutions. Thus, the smallest evaluation value indicates the best focusing solution.

3.4.3.1   Filter Evaluation Criteria

Filter evaluation criteria use statistical characteristics of focusing input and focusing outputs to measure focusing success of focusing solutions. Each filter evaluation criterion considers specific aspects of statistical representativeness. Cumulations of filter evaluation criteria specify overall notions of statistical representativeness. In this sense, these evaluation criteria implement specific focusing criteria for representativeness (see definition 3.2.4, page 50). Table 3.2 presents an overview of filter evaluation criteria and their notations. We define evaluation criteria for modus, mean, variance, distribution, and joint distribution, both in isolated and comparative mode.

Modus and Mean Evaluation in Isolated Mode

We start the definition of filter evaluation criteria with the first specific criterion which takes into account characteristic values in isolated mode. This

criterion considers single attributes and distinguishes between different types of
attributes. In case of qualitative attributes, it compares modus in the focusing
input and modus in focusing outputs. In case of quantitative attributes, this
criterion relates the mean of attribute values in the focusing input to the mean
of attribute values in focusing outputs. Definition 3.4.11 specifies evaluation
criterion $E_{S_j}^{\leftrightarrow}$ for characteristic values in isolated mode.

Definition 3.4.11 (Evaluation Criterion $E_{S_j}^{\leftrightarrow}$)

*Assume a (database) table (A, T), an attribute $a_j \in A$, a focusing input $f_{in} \subseteq T$,
a focusing solution $F : T^{\subseteq} \longrightarrow T^{\subseteq}$, a focusing output $f_{out} = F(f_{in}) \subseteq f_{in}$, if F
is deterministic, or a set of focusing outputs F_{out}, $\forall f_{out} \in F_{out} : f_{out} = F(f_{in}) \subseteq
f_{in}$, if F is non-deterministic, an additional value $\sigma(q_{S_j}) \geq 0$ that depicts the
fraction of the sum of s_{S_j} and its standard deviation and d_{S_j} in case of non-
deterministic focusing solutions and their averaged application, $0 < \alpha < 1$, and
$0 \leq \eta_\sigma \leq 0.5$. We define the evaluation criterion $E_{S_j}^{\leftrightarrow}$ as a function*

$$E_{S_j}^{\leftrightarrow} : F \times T^{\subseteq} \times T^{\subseteq} \longrightarrow [0; \infty[.$$

If a_j qualitative, and

(i)	$:\Longleftrightarrow$	F deterministic	and	$m_j(f_{in}) \equiv m_j(f_{out})$,
(ii)	$:\Longleftrightarrow$	F non-deterministic	and	$m_j(f_{in}) \equiv m_j(f_{out})$

in most cases,

$$E_{S_j}^{\leftrightarrow}(F, f_{in}, f_{out})$$
$$:= \begin{cases} 0, & if\ (i) \\ 1 - \dfrac{|\{f_{out} \in F_{out} \mid m_j(f_{in}) \equiv m_j(f_{out})\}|}{|\{F_{out}\}|}, & if\ (ii) \\ 1, & else \end{cases}$$

If a_j quantitative, and

(i)	$:\Longleftrightarrow$	F deterministic	and	$|q_{S_j}(\alpha, f_{in}, f_{out})| \leq 1$,
(ii)	$:\Longleftrightarrow$	F non-deterministic	and	$|q_{S_j}(\alpha, f_{in}, f_{out})| \leq 1$

in most cases,

(iii)	$:\Longleftrightarrow$	F deterministic	and	$|q_{S_j}(\alpha, f_{in}, f_{out})| > 1$,
(iv)	$:\Longleftrightarrow$	F non-deterministic	and	$|q_{S_j}(\alpha, f_{in}, f_{out})| > 1$

in most cases,

and

$E_{S_j}(F, f_{in}, f_{out})$

$$
:= \begin{cases}
0, & if\ (i) \\[2ex]
1 - \dfrac{\mid \{f_{out} \in F_{out} \ \mid \mid q_{S_j}(\alpha, f_{in}, f_{out}) \mid\ \leq\ 1)\} \mid}{\mid \{F_{out}\} \mid}, & if\ (ii) \\[2ex]
1, & if\ (iii) \\[2ex]
0 + \dfrac{\mid \{f_{out} \in F_{out} \ \mid \mid q_{S_j}(\alpha, f_{in}, f_{out}) \mid\ >\ 1)\} \mid}{\mid \{F_{out}\} \mid}, & if\ (iv)
\end{cases}
$$

$E_{S_j}^{\leftrightarrow}(F, f_{in}, f_{out})$

$$
\begin{aligned}
:= \quad & \tfrac{1}{2} \cdot E_{S_j}(F, f_{in}, f_{out}) \\
+ \quad & \tfrac{1}{2} \cdot \bigg((1 - \eta_\sigma) \cdot \mid q_{S_j}(\alpha, f_{in}, f_{out}) \mid \\
& + \ \eta_\sigma \cdot \mid \sigma(q_{S_j}(\alpha, f_{in}, f_{out})) \mid\ \bigg).
\end{aligned}
$$

In case of qualitative attributes, $E_{S_j}^{\leftrightarrow}$ compares modus in the focusing input and modus in focusing outputs. If the focusing solution is deterministic and both modi coincide, evaluation yields 0. If the modus in the focusing input and the modus in the focusing output do not match (in most cases), it returns 1. Hence, deterministic focusing solutions have high quality, if they correctly represent the modus in the focusing input. In case of non-deterministic focusing solutions, evaluation takes into account the number of times focusing solutions draw focusing outputs with correct modus representations, if the majority of focusing outputs represents the correct modus. In this case, $E_{S_j}^{\leftrightarrow}$ computes the evaluation value as the difference between 1 and the relative number of correct representations. If all focusing outputs represent the same modus as the focusing input, non-deterministic focusing solutions get the same evaluation value as deterministic focusing solutions with correct representation of the modus.

In case of quantitative attributes, evaluation criterion $E_{S_j}^{\leftrightarrow}$ makes use of statistical hypothesis testing and contains two components. The first component is qualitative and measures, if hypothesis testing yields a significant difference between mean in the focusing input and mean in the focusing output. The difference is significant, if and only if the absolute value of the test statistic is larger than the expected distribution. Otherwise, the difference between both means is not significant. We expect that focusing success of focusing solutions which result in focusing outputs with significant differences between their mean and the mean in the focusing input is worse than focusing success of focusing solutions which result in focusing outputs with no significant difference between the respective means. Hence, $E_{S_j}^{\leftrightarrow}$ returns 1 for the first component, if the difference is significant, and 0, if it is not significant. As for qualitative attributes, the first component in $E_{S_j}^{\leftrightarrow}$ for quantitative attributes varies evaluation in case of non-deterministic focusing solutions in a similar vein.

The second component of $E_{S_j}^{\leftrightarrow}$ for quantitative attributes measures the amount of deviation between the mean in the focusing input and its representation in focusing outputs. This amount of deviation corresponds to the relation between the value of the test statistic and the value of the expected distribution. The more the statistic differs from the expected distribution, the higher is the deviation between both means. If the fraction of test statistic and expected distribution is smaller than 1, this fraction estimates the amount of deviation for focusing outputs with non-significant differences between the mean of attribute values in the focusing input and the mean of attribute values in focusing outputs. If this fraction is larger than 1, it indicates the amount of deviation for focusing outputs with significant differences. Note, for non-deterministic focusing solutions, the first part of the second component in $E_{S_j}^{\leftrightarrow}$ holds average values from multiple runs.

The second component in evaluation criterion $E_{S_j}^{\leftrightarrow}$ for quantitative attributes again takes into account, if focusing solutions are deterministic or not. For non-deterministic focusing solutions, $E_{S_j}^{\leftrightarrow}$ adds to the amount of deviation between means the fraction of the sum of average value and standard deviation of the absolute test statistic and the expected distribution. If focusing solutions are non-deterministic, their applications to the same focusing input probably yield different focusing outputs with different qualities. Thus, we suggest to run non-deterministic focusing solutions more than once on the same focusing input and take the average quality of all experiments as the resulting value for evaluation. However, this averaged consideration is optimistic since single runs of non-deterministic focusing solutions possibly get worse results. Thus, we punish the worst case, represented as the sum of average results and standard deviation of all runs, as an additional amount of evaluation criterion $E_{S_j}^{\leftrightarrow}$.

Note, we weigh contributions of deviations between means and an additional amount for non-deterministic focusing solutions according to a pre-specified parameter η_σ. For deterministic focusing solutions, a single experiment is sufficient and the standard deviation is zero. In this case, both parts of the second component in $E_{S_j}^{\leftrightarrow}$ have the same value.

More Filter Evaluation in Isolated Mode

The definitions of isolated filter evaluation criteria for variance, distribution, and joint distribution follow the same structure as the definition of $E_{S_j}^{\leftrightarrow}$. For this reason, we only outline their basic characteristics here, and put their specific definitions into appendix B. Thereby, we avoid less interesting definitions but keep the completeness of all details in order to allow complete re-implementations of all approaches developed in this dissertation.

Definition B.1.1 (see page 267) states filter evaluation which takes into account variances of characteristic values. This evaluation criterion only applies to quantitative attributes. In case of qualitative attributes, we do not have any concept of variance. $E_{V_j}^{\leftrightarrow}$ uses the same template as $E_{S_j}^{\leftrightarrow}$ and also contains two components. The first component measures the qualitative quality of focusing solutions in terms of the significance of differences between variances of attribute values in the focusing input and variances of attribute values in focusing outputs. The second component again considers quantitative information on the amount of difference between the test statistic value and the expected distribution. $E_{V_j}^{\leftrightarrow}$ also punishes non-deterministic focusing solutions as the previous evaluation criterion $E_{S_j}^{\leftrightarrow}$ does. The smaller values of this evaluation criterion are, the more appropriate are representations of variances in the focusing input, and hence, the more appropriate are focusing solutions.

Evaluation criteria $E_D^{\leftrightarrow}$ and $E_J^{\leftrightarrow}$ are similar to evaluation criterion $E_{V_j}^{\leftrightarrow}$ which takes into account variances of statistical characteristics. Definition B.1.2 (see page 268) specifies the evaluation criterion for distributions of values for single attributes, whereas definition B.1.3 (see page 269) establishes the same concept for joint distributions of values for more than single attributes. $E_{D_j}^{\leftrightarrow}$ and $E_J^{\leftrightarrow}$ complete the definition of filter evaluation in isolated mode.

Modus and Mean Evaluation in Comparative Mode

Now, we turn to filter evaluation in comparative mode. Comparative evaluation criteria not only relate the focusing input to results of single focusing solutions but they also compare the focusing input in relation to several focusing outputs. The idea of comparative evaluation is to measure focusing success relatively in comparison to focusing success of other focusing solutions. If focusing solutions do not represent the focusing input well, this is relatively good, if all other focusing outputs do not appropriately represent the focusing input either. As a benchmark for these types of comparisons, we always use the worst possible focusing solution.

The first evaluation criterion $E_{S_j}^{\updownarrow}$ is the comparative variant of $E_{S_j}^{\leftrightarrow}$. It takes into account characteristic values of attributes in the focusing input and in focusing outputs, respectively, and differs between qualitative and quantitative attributes. In case of qualitative attributes, $E_{S_j}^{\updownarrow}$ is identical to $E_{S_j}^{\leftrightarrow}$. This means that we assume that at least one focusing solution in the set of all focusing solutions exists which indeed yields the worst focusing output. Worst focusing outputs do not correctly represent the modus in the focusing input, or all applications of the same non-deterministic focusing solution do not match the modus in the focusing input. Both situations yield an evaluation of 1, which corresponds to the maximum possible value. Thus, the definition of $E_{S_j}^{\updownarrow}$ for qualitative attributes relates focusing success of focusing solutions to the worst focusing output.

Definition 3.4.12 (Evaluation Criterion $E_{S_j}^{\updownarrow}$)

Assume a (database) table (A, T), an attribute $a_j \in A$, a focusing input $f_{in} \subseteq T$, a focusing solution $F : T^{\subseteq} \longrightarrow T^{\subseteq}$, a focusing output $f_{out} = F(f_{in}) \subseteq f_{in}$, if F is deterministic, or a set of focusing outputs F_{out}, $\forall f_{out} \in F_{out} : f_{out} = F(f_{in}) \subseteq f_{in}$, if F is non-deterministic, a set of focusing outputs F'_{out}, $\forall f'_{out} \in F'_{out} : f'_{out} \subseteq f_{in}$, additional values $\sigma(q_{S_j}) \geq 0$ that depict the fraction of the sum of s_{S_j} and its standard deviation and d_{S_j} in case of non-deterministic focusing solutions and their averaged application, $0 < \alpha < 1$, and $0 \leq \eta_\sigma \leq 0.5$. We define the evaluation criterion $E_{S_j}^{\updownarrow}$ as a function

$$E_{S_j}^{\updownarrow} : F \times T^{\subseteq} \times T^{\subseteq} \times \{T^{\subseteq}, \dots, T^{\subseteq}\} \longrightarrow [0; \infty[.$$

If a_j qualitative,

$$E_{S_j}^{\updownarrow}(F, f_{in}, f_{out}, F'_{out}) \quad := \quad E_{S_j}^{\leftrightarrow}(F, f_{in}, f_{out}).$$

If a_j quantitative,

$$E_{S_j}^{\updownarrow}(F, f_{in}, f_{out}, F'_{out})$$

$$:= \quad \tfrac{1}{2} \cdot E_{S_j}(F, f_{in}, f_{out})$$

$$+ \quad \tfrac{1}{2} \cdot \left((1 - \eta_\sigma) \cdot \frac{\mid q_{S_j}(\alpha, f_{in}, f_{out}) \mid}{\max\limits_{f'_{out} \in F'_{out}} \{\mid q_{S_j}(\alpha, f_{in}, f'_{out}) \mid\}} \right.$$

$$+ \quad \eta_\sigma \cdot \left. \frac{\mid \sigma(q_{S_j}(\alpha, f_{in}, f_{out})) \mid}{\max\limits_{f'_{out} \in F'_{out}} \{\mid \sigma(q_{S_j}(\alpha, f_{in}, f'_{out})) \mid\}} \right).$$

The same argumentation is valid for the first component of evaluation criterion $E_{S_j}^{\updownarrow}$ in case of quantitative attributes. Again, the worst focusing output reveals an evaluation of 1, and consequently the first part of the sum in definition 3.4.12 is the relation between results of single focusing solutions and the maximum value among all other focusing outputs. In similar ways, we relate the quantitative amount of deviation between focusing input and focusing output of single focusing solutions to the maximum amount of deviation among all other focusing outputs. Again, evaluation criterion $E_{S_j}^{\updownarrow}$ adds an additional value for non-deterministic focusing solutions and their averaged applications.

More Filter Evaluation in Comparative Mode

We also specify filter evaluation criteria in comparative mode for the other aspects of statistical representativeness. All of them use the same strategy as the previous filter evaluation criterion $E_{S_j}^{\updownarrow}$ to measure focusing success of focusing solutions in relation to focusing success of other focusing solutions. Hence, we again put these definitions into appendix B. Definition B.1.4 (see page 270) states filter evaluation in comparative mode for comparisons of variances of attribute values. Again, this criterion is only applicable for quantitative attributes. Definition B.1.5 (see page 271) is the comparative variant of $E_{D_j}^{\leftrightarrow}$ and considers distributions of attribute values for single attributes, whereas definition B.1.6 (see page 271) constitutes the respective criterion which takes into account joint distributions of attribute values for more than single attributes.

Cumulated Filter Evaluation

Each of the isolated and comparative filter evaluation criteria only investigates single aspects of statistical representativeness. In order to measure focusing success according to an overall notion of statistical representativeness, we now combine these criteria. We propose to define combined evaluation criteria in isolated and comparative modes separately. This allows us to keep absolute and relative qualities of focusing outputs distinct, if we prefer to examine focusing success in both ways.

The following definitions state evaluation criteria which use more than single aspects of statistical representativeness. Both definitions contain additional parameters which weigh each aspect according to its relevance. We either optimize these parameters for each focusing solution separately such that each focusing solution reveals its optimal evaluation result (Nakhaeizadeh & Schnabl, 1997), or we set them according to preferences of domain experts.

Since cumulated filter evaluation criteria in isolated and comparative modes are similar, we only specify cumulated filter evaluation in isolated mode here and move the definition of cumulated filter evaluation in comparative mode to appendix B. Definition 3.4.13 and definition B.1.7 (see page 272) contain four components. The first three components sum evaluation values of the first three evaluation criteria for each attribute. Note, since $E_{V_j}^{\leftrightarrow}$ and $E_{V_j}^{\updownarrow}$ are not applicable to qualitative attributes, this component contains only evaluation values for quantitative attributes. The last term in each of these definitions adds the contribution of evaluation according to joint distributions. We recommend to set η_J to a smaller value than η_S, η_V, and η_D, in order to weigh each evaluation aspect properly.

Definition 3.4.13 (Evaluation Criterion $E_F^{\leftrightarrow}$)

Assume a (database) table (A, T), a focusing input $f_{in} \subseteq T$, a focusing solution $F : T^{\subseteq} \longrightarrow T^{\subseteq}$, a focusing output $f_{out} = F(f_{in}) \subseteq f_{in}$, if F is deterministic, or a set of focusing outputs F_{out}, $\forall f_{out} \in F_{out} : f_{out} = F(f_{in}) \subseteq f_{in}$, if F is non-deterministic, and $0 \leq \eta_S, \eta_V, \eta_D, \eta_J \leq 1$ with $\eta_S + \eta_V + \eta_D + \eta_J = 1$. We define the evaluation criterion $E_F^{\leftrightarrow}$ as a function

$$E_F^{\leftrightarrow} : F \times T^{\subseteq} \times T^{\subseteq} \longrightarrow [0; \infty[,$$

$$E_F^{\leftrightarrow}(F, f_{in}, f_{out})$$

$$
\begin{aligned}
:= \quad & \eta_S \cdot \sum_{j=1}^{N} E_{S_j}^{\leftrightarrow}(F, f_{in}, f_{out}) + \eta_V \cdot \sum_{j=1}^{N} E_{V_j}^{\leftrightarrow}(F, f_{in}, f_{out}) \\
+ \quad & \eta_D \cdot \sum_{j=1}^{N} E_{D_j}^{\leftrightarrow}(F, f_{in}, f_{out}) + \eta_J \cdot E_J^{\leftrightarrow}(F, f_{in}, f_{out}).
\end{aligned}
$$

At the end of the definitions of filter evaluation criteria, we define a final filter evaluation criterion which combines cumulated evaluations in isolated and comparative modes. This criterion abstracts from all detailed evaluation information and contains all aspects of statistical representativeness from absolute and relative comparisons. Again, we use two additional parameters to weigh the amount of contributions of $E_F^{\leftrightarrow}$ and $E_F^{\updownarrow}$. We suggest to regard absolute comparisons as less relevant than relative considerations. Hence, we argue to set $\eta^{\leftrightarrow}$ to a smaller value than $\eta^{\updownarrow}$. Definition 3.4.14 consolidates all filter evaluation criteria and establishes a compact measure for focusing success according to statistical representativeness.

Definition 3.4.14 (Evaluation Criterion E_F)

Assume a (database) table (A, T), a focusing input $f_{in} \subseteq T$, a focusing solution $F : T^{\subseteq} \longrightarrow T^{\subseteq}$, a focusing output $f_{out} = F(f_{in}) \subseteq f_{in}$, if F is deterministic, or a set of focusing outputs F_{out}, $\forall f_{out} \in F_{out} : f_{out} = F(f_{in}) \subseteq f_{in}$, if F is non-deterministic, a set of focusing outputs F'_{out}, $\forall f'_{out} \in F'_{out} : f'_{out} \subseteq f_{in}$, and $0 \leq \eta^{\leftrightarrow}, \eta^{\updownarrow} \leq 1$. We define the evaluation criterion E_F as a function

$$E_F : F \times T^{\subseteq} \times T^{\subseteq} \times \{T^{\subseteq}, \dots, T^{\subseteq}\} \longrightarrow [0; \infty[,$$

$$E_F(F, f_{in}, f_{out}, F'_{out})$$

$$:= \eta^{\leftrightarrow} \cdot E_F^{\leftrightarrow}(F, f_{in}, f_{out}) + \eta^{\updownarrow} \cdot E_F^{\updownarrow}(F, f_{in}, f_{out}, F'_{out}).$$

3.4.3.2 Wrapper Evaluation Criteria

Recall, the most important aspects in wrapper evaluation are execution time, storage requirements, accuracy, and complexity of data mining results. In the following, we define isolated and comparative evaluation criteria for each of these aspects. Again, we combine these criteria to cumulated isolated and cumulated comparative evaluation criteria, respectively. Finally, we specify an overall wrapper evaluation criterion which puts together isolated and comparative criteria. Table 3.3 presents an overview of isolated and comparative wrapper evaluation criteria and their notations.

Table 3.3: Isolated and Comparative Wrapper Evaluation Criteria

	Execution Time	Storage Requirements	Accuracy	Complexity
isolated	$E_T^{\leftrightarrow}$	$E_M^{\leftrightarrow}$	$E_A^{\leftrightarrow}$	$E_C^{\leftrightarrow}$
comparative	$E_T^{\updownarrow}$	$E_M^{\updownarrow}$	$E_A^{\updownarrow}$	$E_C^{\updownarrow}$

Execution Time Evaluation in Isolated Mode

The first isolated wrapper evaluation criterion $E_T^{\leftrightarrow}$ takes into account execution time of focusing and data mining. It relates the focusing input to single focusing solutions and their focusing outputs. We are interested in comparisons of execution time of focusing on the focusing input and data mining on focusing outputs against data mining on the focusing input. If we only evaluate focusing solutions according to execution time, we expect that for appropriate focusing solutions the sum of execution time of focusing and data mining is smaller than execution time of data mining on the entire focusing input. Consequently, we define the evaluation criterion which takes into account execution time as a relation between the sum of execution time of focusing and data mining and execution time of data mining on the focusing input. Definition 3.4.15 specifies the respective evaluation criterion $E_T^{\leftrightarrow}$.

Definition 3.4.15 (Evaluation Criterion $E_T^{\leftrightarrow}$)

Assume a (database) table (A, T), a focusing input $f_{in} \subseteq T$, a focusing solution $F : T^{\subseteq} \longrightarrow T^{\subseteq}$, a focusing output $f_{out} = F(f_{in}) \subseteq f_{in}$, if F is deterministic, or a set of focusing outputs F_{out}, $\forall f_{out} \in F_{out} : f_{out} = F(f_{in}) \subseteq f_{in}$, if F is non-deterministic, a data mining algorithm D, an additional value $\sigma(t) \geq 0$ that depicts the sum of t and its standard deviation in case of non-deterministic focusing solutions and their averaged application, and $0 \leq \eta_\sigma \leq 0.5$. We define the evaluation criterion $E_T^{\leftrightarrow}$ as a function

$$E_T^{\leftrightarrow} : F \times D \times T^{\subseteq} \times T^{\subseteq} \longrightarrow]0; \infty[,$$

$$E_T^{\leftrightarrow}(F, D, f_{in}, f_{out})$$
$$:= (1 - \eta_\sigma) \cdot \left(\frac{t(F, f_{in}) + t(D, f_{out})}{t(D, f_{in})} \right)$$
$$+ \eta_\sigma \cdot \left(\frac{\sigma\left(t(F, f_{in})\right) + \sigma\left(t(D, f_{out})\right)}{t(D, f_{in})} \right).$$

The first part of $E_T^{\leftrightarrow}$ returns values smaller than 1, if the sum of execution time of focusing on the focusing input and data mining on the focusing output is smaller than execution time of data mining on the focusing input. It reveals values larger than 1, if the inverse is true. If we compare focusing solutions according to this evaluation criterion, more appropriate focusing solutions with respect to execution time yield smaller values of $E_T^{\leftrightarrow}$.

Note, in case of non-deterministic focusing solutions, f_{out} in definition 3.4.15 refers to average applications of the same focusing solution. For simplicity, we do not explicitly compute average values of execution time but rather assume that t already covers average values for all focusing outputs in F_{out}. We use the same simplification in all of the following definitions, too.

The second term in $E_T^{\leftrightarrow}$ again punishes non-deterministic focusing solutions. Recall, if focusing solutions are non-deterministic, their applications to the same focusing input probably yield different focusing outputs with different qualities. Thus, we take the average quality of multiple experiments as the evaluation result and punish the worst case as an additional amount of evaluation criterion $E_T^{\leftrightarrow}$. For deterministic focusing solutions, both parts of $E_T^{\leftrightarrow}$ have the same value.

More Wrapper Evaluation in Isolated Mode

The definition of isolated wrapper evaluation criteria for storage requirements, accuracy, and complexity follow the same structure as the definition of $E_{T_j}^{\leftrightarrow}$. Thus, we only outline their basic characteristics here and put their specific definitions into appendix B.

The second isolated wrapper evaluation criterion $E_M^{\leftrightarrow}$ takes into account storage requirements. It relates the number of tuples in the focusing input to focusing output sizes. The smaller the focusing outputs are, the better is the quality of respective focusing solutions. We define $E_M^{\leftrightarrow}$ as the fraction of focusing output size and focusing input size (see definition B.2.1, page 272). Again, we punish the worst case in terms of the sum of averaged focusing output sizes over multiple runs and their standard deviations for non-deterministic focusing solutions. $E_M^{\leftrightarrow}$ always returns positive values. The larger the value is, the lower is the quality of the focusing solution.

Note, in contrast to the first evaluation criterion according to execution time, we do not consider the sum of storage requirements of focusing and data mining. Since storage requirements of focusing solutions are always the same in terms of the number of tuples, relations among evaluations of different focusing solutions remain constant. Although this is an idealized assumption about storage requirements of focusing solutions, we focus our attention on storage requirements of data mining.

The third aspect in isolated wrapper evaluation is accuracy of classifiers. Evaluation criterion $E_A^{\leftrightarrow}$ compares error rates of classifiers which result from applications of data mining to the entire focusing input and from applications of data mining to focusing outputs (see definition B.2.2, page 273). The construction of $E_A^{\leftrightarrow}$ is similar to the definition of the first evaluation criterion. Since we usually expect increasing error rates for classifiers generated on focusing outputs, we divide their error rates by error rates of classifiers from the focusing input.[6] Again, we punish non-deterministic focusing solutions in the same way as in the definition of the first two evaluation criteria.

The last isolated wrapper evaluation criterion $E_C^{\leftrightarrow}$ judges complexity of data mining results. Definition B.2.3 (see page 273) is similar to the second evaluation criterion $E_M^{\leftrightarrow}$. Again, we relate complexity of data mining results on focusing outputs to complexity of data mining results on the entire focusing input. Since complexity in case of top down induction of decision trees contains two components, number of levels and number of nodes in decision trees, we combine both aspects into the definition of this evaluation criterion. If the data mining algorithm is a nearest neighbor classifier, we identify complexity of data mining results with storage requirements. Like the previous evaluation criteria, $E_C^{\leftrightarrow}$ also contains an additional term to punish non-deterministic focusing solutions.

Execution Time Evaluation in Comparative Mode

Now, we also specify wrapper evaluation in comparative mode. Since these criteria in comparative mode follow up the structure of filter evaluation criteria in comparative mode, we only define an example criterion according to execution time here and refer to appendix B for the remaining wrapper evaluation criteria in comparative mode.

Definition 3.4.16 specifies wrapper evaluation in comparative mode according to execution time. Whereas its isolated variant relates execution time of focusing on the focusing input and data mining on focusing outputs to execution time of data mining on the focusing input, $E_T^{\updownarrow}$ now compares the sum of execution time of focusing and data mining for different focusing solutions. Again, we use the maximum sum among all focusing solutions as the benchmark which represents the worst focusing output. This criterion enables relative comparisons which judge focusing success relatively well, if most of the other focusing solutions yield worse results. As in previous definitions, evaluation criterion $E_T^{\updownarrow}$ punishes non-deterministic focusing solutions by an additional amount, in order to reflect their potentials for drawing particular bad focusing outputs in single runs.

Definition 3.4.16 (Evaluation Criterion $E_T^{\updownarrow}$)

Assume a (database) table (A, T), a focusing input $f_{in} \subseteq T$, a focusing solution $F : T^{\subseteq} \longrightarrow T^{\subseteq}$, a focusing output $f_{out} = F(f_{in}) \subseteq f_{in}$, if F is deterministic, or a set of focusing outputs F_{out}, $\forall f_{out} \in F_{out} : f_{out} = F(f_{in}) \subseteq f_{in}$, if F is

---

[6]In order to avoid division by zero, if the error rate of data mining results on the entire focusing input is zero, we add a small value to both terms of the fraction in these situations.

non-deterministic, a set of focusing outputs F'_{out}, $\forall f'_{out} \in F'_{out} : f'_{out} \subseteq f_{in}$, a data mining algorithm D, an additional value $\sigma(t) \geq 0$ that depicts the sum of t and its standard deviation in case of non-deterministic focusing solutions and their averaged application, and $0 \leq \eta_\sigma \leq 0.5$. We define the evaluation criterion $E_T^\updownarrow$ as a function

$$E_T^\updownarrow : F \times D \times T^\subseteq \times T^\subseteq \times \{T^\subseteq, \dots, T^\subseteq\} \longrightarrow]0; 1],$$

$$E_T^\updownarrow(F, D, f_{in}, f_{out}, F'_{out})$$

$$:= (1 - \eta_\sigma) \cdot \frac{t(F, f_{in}) + t(D, f_{out})}{\max\limits_{f'_{out} \in F'_{out}} \{t(F', f_{in}) + t(D, f'_{out})\}}$$

$$+ \eta_\sigma \cdot \frac{\sigma(t(F, f_{in})) + \sigma(t(D, f_{out}))}{\max\limits_{f'_{out} \in F'_{out}} \{\sigma(t(F', f_{in})) + \sigma(t(D, f'_{out}))\}}.$$

More Wrapper Evaluation in Comparative Mode

The remaining three wrapper evaluation criteria in comparative mode according to storage requirements, accuracy, and complexity of data mining results follow up the idea of evaluation criterion $E_T^\updownarrow$. We always relate focusing success of single focusing solutions to focusing success of other focusing solutions. Again, we represent the benchmark of other focusing solutions by the maximum value of each aspect which refers to worst focusing outputs. Each comparative wrapper evaluation criteria considers the same aspect as their isolated counterparts. In appendix B, we precisely define comparative wrapper evaluation criteria for storage requirements (see definition B.2.4, page 274), accuracy (see definition B.2.5, page 274), and complexity (see definition B.2.6, page 275). All definitions punish non-deterministic focusing solutions in the same way as before.

Cumulated Wrapper Evaluation

As in filter evaluation, we also want to measure focusing success according to more than single aspects in wrapper evaluation. Hence, we define cumulated isolated as well as cumulated comparative wrapper evaluation criteria. Both criteria again sum up all aspects in isolated and comparative modes, respectively. Definition 3.4.17 specifies $E_W^\leftrightarrow$ as an example of isolated cumulated wrapper evaluation, whereas we put its comparative variant $E_W^\updownarrow$ into appendix B (see definition B.2.7, page 275), since the patterns of both criteria are the same. Again, these definitions contain four additional parameters to weigh each aspect in wrapper evaluation properly. We recommend to set η_A to a higher value than η_T, η_M, and η_C since accuracy is usually the most important aspect in wrapper evaluation.

Definition 3.4.17 (Evaluation Criterion $E_W^{\leftrightarrow}$)

Assume a (database) table (A, T), a focusing input $f_{in} \subseteq T$, a focusing solution $F : T^{\subseteq} \longrightarrow T^{\subseteq}$, a focusing output $f_{out} = F(f_{in}) \subseteq f_{in}$, if F is deterministic, or a set of focusing outputs F_{out}, $\forall f_{out} \in F_{out} : f_{out} = F(f_{in}) \subseteq f_{in}$, if F is non-deterministic, a data mining algorithm D, a test set $T_{test} \subseteq T$, and $0 \leq \eta_T, \eta_M, \eta_A, \eta_C \leq 1$ with $\eta_T + \eta_M + \eta_A + \eta_C = 1$. We define the evaluation criterion $E_W^{\leftrightarrow}$ as a function

$$E_W^{\leftrightarrow} : F \times D \times T^{\subseteq} \times T^{\subseteq} \times T^{\subseteq} \longrightarrow \ ]0; \ \infty[,$$

$$E_W^{\leftrightarrow}(F, D, f_{in}, f_{out}, T_{test})$$

$$:= \quad \eta_T \cdot E_T^{\leftrightarrow}(F, D, f_{in}, f_{out}) \ + \ \eta_M \cdot E_M^{\leftrightarrow}(D, f_{in}, f_{out})$$

$$+ \quad \eta_A \cdot E_A^{\leftrightarrow}(D, f_{in}, f_{out}, T_{test}) \ + \ \eta_C \cdot E_C^{\leftrightarrow}(D, f_{in}, f_{out}).$$

In summary, we now define a consolidated wrapper evaluation criterion which takes into account all aspects in isolated and comparative modes. The overall wrapper evaluation criterion E_W also contains two additional parameters in order to weigh the relevance of absolute and relative focusing success. Again, we recommend to set $\eta^{\leftrightarrow}$ to a smaller value than $\eta^{\updownarrow}$ since we regard relative focusing success as more important than absolute focusing success. Definition 3.4.18 specifies the final wrapper evaluation criterion for analyses of focusing success in terms of the quality of data mining results.

Definition 3.4.18 (Evaluation Criterion E_W)

Assume a (database) table (A, T), a focusing input $f_{in} \subseteq T$, a focusing solution $F : T^{\subseteq} \longrightarrow T^{\subseteq}$, a focusing output $f_{out} = F(f_{in}) \subseteq f_{in}$, if F is deterministic, or a set of focusing outputs F_{out}, $\forall f_{out} \in F_{out} : f_{out} = F(f_{in}) \subseteq f_{in}$, if F is non-deterministic, a set of focusing outputs F'_{out}, $\forall f'_{out} \in F'_{out} : f'_{out} \subseteq f_{in}$, a data mining algorithm D, a test set $T_{test} \subseteq T$, and $0 \leq \eta^{\leftrightarrow}, \eta^{\updownarrow} \leq 1$. We define the evaluation criterion E_W as a function

$$E_W : F \times D \times T^{\subseteq} \times T^{\subseteq} \times \{T^{\subseteq}, \dots, T^{\subseteq}\} \times T^{\subseteq} \longrightarrow \ ]0; \ \infty[,$$

$$E_W(F, D, f_{in}, f_{out}, F'_{out}, T_{test})$$

$$:= \eta^{\leftrightarrow} \cdot E_W^{\leftrightarrow}(F, D, f_{in}, f_{out}, T_{test}) \ + \ \eta^{\updownarrow} \cdot E_W^{\updownarrow}(F, D, f_{in}, f_{out}, F'_{out}, T_{test}).$$

3.5 Selecting the Focusing Task

In summary, this chapter defined prerequisites for formal specifications of focusing tasks. We discussed focusing input, focusing output, and focusing criterion as components of focusing specifications. We also characterized the focusing context and its impact on focusing. Data mining goal, data characteristics, and data mining algorithms constitute the focusing context. Each aspect influences the definition of focusing tasks as well as the development or selection of appropriate focusing solutions. Finally, we also addressed different evaluation strategies in order to analyze focusing success of focusing solutions. We defined quantitative filter and wrapper evaluation criteria in isolated and comparative modes as well as cumulated measures which take into account more than single evaluation aspects. Whereas filter evaluation emphasizes aspects of statistical representativeness, wrapper evaluation measures focusing success in terms of the quality of data mining results.

Definition 3.5.1 (Focusing Task)

Assume a (database) table (A, T), a focusing input f_{in}, a focusing output f_{out}, a focusing criterion $f(f_{in}, f_{out})$, a data mining goal G, data characteristics $C(A, f_{in})$, and a data mining algorithm D. We define a focusing task as a tuple $(f_{in}, f_{out}, f(f_{in}, f_{out}), G, C(A, f_{in}), D)$.

In conclusion, *focusing tasks* consist of focusing specification and focusing context. Definition 3.5.1 specifies focusing tasks and consolidates the contents of the previous and this chapter and their definitions. As stated previously, definitions of focusing tasks map the intuitive meaning of focusing to formal descriptions. The formal definition of focusing tasks specifies requirements for successful focusing solutions.

At this point, we specialize the scope of work in this dissertation again. At the end of chapter 2, we already fixed the goal of this work to development, analysis, and evaluation of focusing solutions for classification goals and two specific data mining algorithms, top down induction of decision trees and nearest neighbor classifiers. Now, we narrow this scope to specific focusing tasks with the following characteristics:

- *Focusing Input*

  The focusing input is a subset of tuples, i.e., $f_{in} \subseteq T$.

- *Focusing Output*

  The focusing output is a simple subset or a constrained subset, i.e., $f_{out} \subseteq f_{in}$, or $f_{out} \subseteq f_{in}$ and $P(f_{in}, f_{out})$.

- *Focusing Criterion*

  The focusing criterion $f(f_{in}, f_{out})$ is representativeness.

- *Data Mining Goal*

  The data mining goal G is classification.

- *Data Characteristics*

  We do not restrict data characteristics $C(A, f_{in})$ except that the set of attributes contains a qualitative class attribute $a_N \in A$.

- *Data Mining Algorithms*

  The data mining algorithm D is either top down induction of decision trees or nearest neighbor.

Focusing solutions are implementations of algorithms which generate focusing outputs on the focusing input. Evaluation criteria implement focusing criteria within the focusing context and measure focusing success of focusing solutions. Definition 3.5.2 specifies this notion of focusing solutions including the optimization of evaluation criteria.

Definition 3.5.2 (Focusing Solution)

Assume a (database) table (A, T), a focusing task $(f_{in}, f_{out}, f(f_{in}, f_{out}), G, C(A, f_{in}), D)$, and an evaluation criterion E. We define a focusing solution F as an implementation of an algorithm $F : T^{\subseteq} \longrightarrow T^{\subseteq}$ with $F(f_{in}) = f_{out}$. The focusing success of F is the optimization of E.

We further restrict development, analysis, and evaluation of focusing solutions to implementations of algorithms which attempt to optimize the specific evaluation criteria E_F and E_W:

- *Evaluation Criterion*

  The evaluation criterion E is E_F or E_W (see definition 3.4.14, page 77, and definition 3.4.18, page, 82, respectively).

In the following chapters, we develop, analyze, and evaluate focusing solutions according to these specializations of focusing tasks. These specializations complete the precise definition of the primary goal of this dissertation.[7]

---

[7]Note, although we restrict our attention to focusing tuples, we indirectly deal with focusing attributes and focusing values, too.

Chapter 4

Focusing Solutions

In this chapter, we first analyze the state of the art of existing approaches towards focusing solutions and discuss their advantages and disadvantages. We present a unifying framework that covers all existing approaches as instantiations. Reuse of most promising efforts and enhancements to the unifying framework result in a generic algorithm for more intelligent sampling. Its implementation in a commercial data mining tool allows easy applications of different focusing solutions and straightforward integrations of focusing solutions into KDD processes. The primary goal of this chapter is the development of focusing solutions based on existing reusable components.

4.1 State of the Art: A Unifying View

We concentrate the review of state of the art focusing solutions and related efforts on statistics and machine learning approaches. We observe that all existing approaches use statistical sampling techniques or rely on specific approaches to clustering accompanied by an additional prototyping step. Some existing efforts enhance these basic techniques, others combine more than a single technique. In this section, we define a unifying framework that covers all existing approaches towards focusing solutions as instantiations. Then, we present example instantiations, characterize their contributions to focusing solutions, and analyze their strengths and weaknesses. For the development of more intelligent sampling techniques, we reuse their positive aspects whereas we overcome their drawbacks.

4.1.1 The Unifying Framework of Existing Focusing Solutions

The general assumption of the unifying framework is either to utilize statistical sampling techniques, or to search for sets of sufficiently similar tuples in

T. Reinartz: Focusing Solutions for Data Mining, LNAI 1623, pp. 85-158, 1999
© Springer-Verlag Berlin Heidelberg 1999

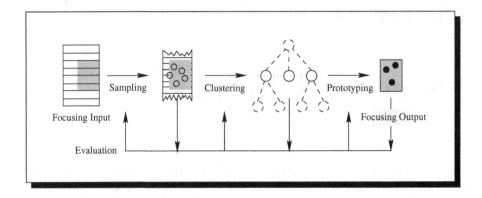

Figure 4.1: The Unifying Framework of Existing Focusing Solutions

the focusing input and to replace each set by a single or a few prototypes. Instantiations of this framework differ in their way to apply statistical sampling techniques, clustering methods, and prototyping approaches. Figure 4.1 outlines the unifying framework that consists of three steps:

1. *Sampling*

2. *Clustering*

3. *Prototyping*

For instance, a generic example instantiation of this framework works as follows. First, an application of a statistical sampling technique draws an initial sample. In the next step, a clustering technique groups the initial sample into subsets of similar tuples. For each of these subsets, the prototyping step selects or constructs a smaller set of representative prototypes. If the clustering step uses hierarchical clustering techniques (see section 4.1.3, page 99), selection of a most appropriate level in the resulting hierarchy precedes prototyping.

At any intermediate point, focusing solutions may apply an evaluation step in order to judge whether the focusing output already meets the pre-defined evaluation criterion. In this framework, evaluation criteria are as simple as checking the sample size, or as complex as applying data mining algorithms and analyzing their results. In addition to evaluation criteria for focusing success, each intermediate step possibly contains specific local evaluation criteria. For example, in case of hierarchical clustering, instantiations of this framework often apply criteria to measure the appropriateness of cluster structures at different levels in the hierarchy.

The order of steps in this framework is not strict. Instantiations of this framework apply the basic steps in any order, or skip some steps completely.

For example, applications of *simple random sampling* also result in instantiations of this framework, and hence they provide particular focusing solutions. Similarly, instantiations sometimes also skip one of the basic steps, perform other steps first, and then return to the previously ignored step. For example, *cluster sampling* is an instantiation of this framework that first applies clustering techniques and then uses sampling to draw the final focusing output. Other approaches combine more than a single step into a compound step. For example, *non-hierarchical clustering techniques* often build groups of similar tuples with representatives as the seeds of clusters. These representatives are specific types of prototypes. Hence, such approaches perform clustering and prototyping in a single procedure.

In the following subsections, we describe sampling, clustering, and prototyping in more detail and discuss example instantiations of this framework. Thereby, we concentrate descriptions of existing solutions on the specific step in the unifying framework that the respective solution mainly applies.

4.1.2   Sampling

The first step of the unifying framework uses statistical *sampling* techniques (Cochran, 1977; Scheaffer *et al.*, 1996). In general, statistical sampling techniques mostly supply surveys in human populations. Starting with an initial question which requires measurements of variables in the entire population, statistical sampling techniques lead to approximate answers by drawing samples from the population, measuring variables of interest in these samples, and then concluding from results on samples to the entire population. Typically, variables of interest are numeric values such as averages, sums, or proportions. Although KDD emphasizes on data mining, i.e., applications of automatic techniques to extract more complex information from data, the primary goal of data reduction in order to enable analyses of data is similar in statistical sampling and focusing.

As a general advantage of statistical sampling techniques, we recognize that most sampling techniques are efficient in terms of running time, and from KDD perspectives costs for drawing samples from (database) tables and calculating statistical values are low. The theory of statistical sampling is also well understood, although it only applies to estimations of a few numeric statistical values, and a similar theory for the appropriateness of statistical sampling techniques in KDD does not exist. We also notice that statistical sampling techniques generally do not take into account the focusing context and represent non-deterministic focusing solutions. Hence, there are always fractions of uncertainty in practice, and single experiments using statistical sampling techniques are not sufficient for KDD purposes.

Since most existing focusing solutions use derivatives of statistical sampling techniques, and we reuse statistical sampling techniques as an auxiliary instrument in more intelligent sampling, we now review efforts towards focusing solutions in statistics and present their most important characteristics. At the end of the discussion on statistical sampling techniques, we also describe two approaches that utilize these techniques in the context of KDD.

4.1.2.1 Simple Random Sampling

Simple random sampling selects tuples from the focusing input such that each possible distinct focusing output of the same size has equal probability. In practice, we draw simple random samples tuple by tuple. If we assume that the focusing input contains M tuples, we number tuples in the focusing input from 1 to M and draw m random numbers between 1 and M, if the desired focusing output size is m. We draw random numbers either by means of a table which lists random numbers or by means of a computer program that produces such a table. The tuples which correspond to the selected m numbers constitute the final sample. At any draw, this process must give an equal chance of selection to any tuple in the focusing input.

Algorithm 4.1.1   RanSam(f_{in}, m)

---

begin
  $f_{out} := \emptyset$;
  while $\mid f_{out} \mid \leq m$ **do**
        $i := random(1, \mid f_{in} \mid)$;
        $f_{out} := f_{out} \cup \{t_i\}$;
  enddo
  return(f_{out});
end;

Algorithm 4.1.1 shows an implementation of simple random sampling. After initialization of the focusing output to an empty set, this algorithm iteratively uses function *random* to generate a random number between 1 and the size of the focusing input. The tuple with this number as its index in the focusing input is then added to the focusing output. Random sampling iterates this process until the focusing output contains a pre-specified number m of tuples.

Note, in RanSam we use simple random sampling *with replacement*, i.e., each number has the same chance at each draw regardless whether it has already been sampled or not. If we exclude drawn numbers from further selection, the resulting sampling strategy is called simple random sampling *without replacement*. The theory of sampling provides simpler formulas for variances and estimated variances of approximations for statistical values in case of sampling with replacement than for sampling without replacement. For this reason, statistics prefer sampling with replacement in more complex sampling surveys. From a technical point of view, we also need less memory capacities, if we draw random samples with replacement, since we do not need to record which tuples have already been selected.

4.1.2.2 Systematic Sampling

Systematic sampling is different from simple random sampling. Again, we assume that the focusing input of size M is numbered from 1 to M. For a pre-specified number *step*, systematic sampling draws the first tuple out of the first *step* tuples in the focusing input at a random position. Then, it iteratively adds each tuple with an index which refers to *step* positions after the selection position in the previous step. For example, if the first tuple in the focusing output is the third tuple out of seven in the focusing input, subsequent selections are the tenth tuple, the 17th tuple, the 24th tuple, and so on. In systematic sampling, random selection of the first tuple, specification of the step size, and the order of tuples in the focusing input determine the entire sample.

If we apply systematic sampling as described above, the focusing output size inherently depends on the step size. If the focusing input contains M tuples and the step size is *step*, the final sample size is $\lfloor \frac{M}{step} \rfloor$.[1] If we want to specify the sample size in advance and then use systematic sampling to draw a focusing output of this size, we use the following alternative. Assume, we specify a sample size m. Then, we compute the step size as $step := \lfloor \frac{M}{m} \rfloor$ and apply systematic sampling as described above. This strategy ensures that the focusing output is equally spread over the entire focusing input.

Algorithm 4.1.2   SYSSAM($f_{in}, start, m$)

---

begin
   $f_{out} := \emptyset$;
   $step := \lfloor \frac{|f_{in}|}{m} \rfloor$;
   $i := start$;
   while $i \leq |f_{in}|$ **do**
        $f_{out} := f_{out} \cup \{t_i\}$;
        $i := i + step$;
   enddo
   return(f_{out});
end;

Algorithm 4.1.2 describes an implementation of systematic sampling which uses this alternative to compute the step size and also allows an explicit specification of the *start* position in contrast to systematic sampling in statistics. After initialization of the focusing output to an empty set, SYSSAM computes the step size according to the number of tuples in the focusing input and the required focusing output size m. Then, it sets the initial selection position to the given *start* value and continues adding tuples to the focusing output until the incremented index is larger than the number of tuples in the focusing input. If we fix the start position and define a well-defined order on the focusing input, SYSSAM represents a deterministic version of systematic sampling.

---

[1] $\lfloor \frac{M}{step} \rfloor$ depicts the largest integer value smaller than $\frac{M}{step}$.

There are two apparent advantages of systematic sampling in comparison to simple random sampling. First, it is easier to select systematic samples without computation of random numbers at each draw, and it is often also easier to execute systematic sampling without mistakes since we do not need to ensure equal probabilities for each possible focusing output of the same size. Second, systematic sampling is likely to be more precise than simple random sampling since systematic samples are automatically spread more evenly across the population. If the focusing input is ordered and its size is large, systematic sampling is preferable in comparison to simple random sampling, but if the order of tuples is random, results of both sampling techniques are comparable. On the other hand, random sampling is more reliable than systematic sampling, if there is a periodic tendency in the focusing input.

4.1.2.3 Stratified Sampling

Stratified sampling extends simple random sampling and systematic sampling by an additional preparation step. Stratified sampling first separates the focusing input into subsets. Each of these subsets is a *stratum*, and the resulting set of strata builds non-overlapping and exhaustive partitions of the original focusing input. After construction of strata, this sampling strategy draws samples from each stratum independently. If we draw each sample by an application of simple random sampling, the resulting procedure is *stratified simple random sampling*. Instead of using simple random sampling within each stratum, we can also use any other sampling technique.

Algorithm 4.1.3 $\mathrm{STRSAM}(f_{in}, m)$

---

begin
   $f_{out} := \emptyset;$
   $S := stratify(f_{in}, ...);$
   $l := 1;$
   while $l \leq |S|$ **do**
      $m_l := \lfloor \frac{|s_l|}{|f_{in}|} \cdot m \rfloor + 1;$
      $f_{out} := f_{out} \cup \mathrm{RANSAM}(s_l, m_l);$
      $l := l + 1;$
   enddo
   return$(f_{out});$
end;

Algorithm 4.1.3 shows an implementation of stratified simple random sampling. After initialization of the focusing output to an empty set, STRSAM separates the focusing input into strata $S := \{s_1, \ldots, s_l, \ldots, s_L\}$ by utilizing procedure $stratify.$[2] Then, stratified sampling iteratively draws simple ran-

---

[2]Note, the number of parameters in procedure $stratify$ depends on the specific

dom samples within each stratum. Therefore, STRSAM computes the sample size within each stratum proportional to the stratum size. This strategy is usually appropriate, although optimum allocation is different (see Scheaffer *et al.* (1996), page 141). However, optimum allocation requires more computational effort since it uses variances within each stratum, and variances are hard to measure in the context of focusing solutions since we do not simply compute means or sums. At the end, STRSAM returns the unification of all samples as the final focusing output.

Note, STRSAM slightly differs from stratified simple random sampling in statistics in the following way. Whereas we unify samples within each stratum to an overall sample and then use this sample for further processing, the original version in statistics first estimates statistical values within each stratum and then combines these separate estimates to an overall estimation of the desired parameter.

Beyond practical reasons in sampling surveys, the main advantage of stratified sampling is the separation of more heterogeneous focusing inputs into more homogeneous strata. Consequently, small samples within each stratum yield precise estimates of characteristics in the entire stratum, and combinations of all estimates result in precise estimates for the entire focusing input. In comparison to simple random sampling, stratified sampling yields lower variances, if we estimate proportions. This gain in precision is particularly large, if proportions vary from stratum to stratum (see Cochran (1977), page 109).

Stratify

The theory of stratified sampling usually deals with properties of estimates from stratified samples and with the best choice of sample sizes within each stratum in order to obtain maximum precision. Construction of strata and descriptions of alternatives for procedure *stratify* in algorithm 4.1.3 are discussed in less detail. We present a specific stratification procedure in section 4.2.2.4.

Stratified Sampling and Systematic Sampling

There is a close relation between stratified sampling and systematic sampling. In some sense, systematic sampling is a special case of stratified sampling. If we draw systematic samples with step size *step*, we consider each set of *step* subsequent tuples in the focusing input as a single stratum. The first stratum contains the first *step* tuples, the second stratum contains the second *step* tuples, and so on. Systematic sampling then chooses a single tuple within each stratum starting at the same random or the same pre-specified start position. The unification of all tuples corresponds to a stratified sample. The difference between stratified sampling and systematic sampling is that STRSAM selects tuples at random positions within each stratum whereas SYSSAM always draws tuples at the same relative position.

---

mechanism to separate the focusing input into strata.

Stratified Sampling and Cluster Sampling

Similar to stratified sampling, *cluster sampling* assumes a separation of the focusing input into subsets of tuples, but the assumptions on these subsets are different. Whereas strata represent homogeneous subsets and unifications of samples across all strata constitute representative focusing outputs, each cluster in cluster sampling reveals heterogeneous representations of the focusing input, and selections of all tuples in a few clusters constitute cluster samples. Note, this type of clusters in cluster sampling is different from clusters that result from the second step in the unifying framework. Clustering algorithms separate the focusing input into subsets of similar tuples. Hence, homogeneity within clusters is usually higher than homogeneity between clusters which is not true in case of cluster sampling.

4.1.2.4 Windowing

Most of existing efforts towards focusing solutions in KDD utilize statistical sampling techniques, and we now outline two specific example approaches. The first example is *windowing* (Quinlan, 1986), a modification of top down induction of decision trees (see section 2.5.2, page 33). Windowing allows to apply focusing solutions during generation of classifiers. The resulting modified version of algorithm TDIDT uses samples of increasing sizes to iteratively improve an existing classifier.

Initially, windowing draws a simple random sample from the focusing input, builds a classifier on this sample using top down induction of decision trees, and applies the resulting classifier to remaining tuples in the focusing input. Then, windowing enlarges the initial training set by a random selection of misclassified tuples. Top down induction of decision trees creates a new classifier on this enlarged training set and again the remaining set of tuples not included in the current window constitutes the test set. Windowing repeats this process until the resulting decision tree classifies all remaining tuples correctly. Usually, the final window ends up containing only a fraction of the original focusing input. The final window is thus a representative sample of the focusing input.

Tuple Selection in Windowing

In terms of the framework for the definition of focusing tasks, focusing outputs in windowing are constrained subsets of the focusing input. Definition 4.1.1 states the selection criterion in windowing. This criterion evaluates to true at the end of windowing, if we consider the final window as the focusing output and the classifier which results from an application of algorithm TDIDT to this focusing output classifies each remaining tuple in the focusing input correctly. Note, the selection process in windowing is not deterministic but applies simple random sampling. In this sense, the selection criterion is used to evaluate the current focusing output but not to generate the focusing output.

Definition 4.1.1 (Tuple Selection in Windowing)

Assume a (database) table (A, T), a focusing input $f_{in} \subseteq T$, a focusing output $f_{out} \subseteq f_{in}$, and a top down induction of decision trees algorithm $TDIDT$. Predicate $P(f_{in}, f_{out})$ in windowing is

$$P(f_{in}, f_{out}) \quad :\Longleftrightarrow \quad \forall t_i \in f_{in} - f_{out} : \; TDIDT(f_{out})(t_{i[1; \, N-1]}) \equiv t_{iN}.$$

Enhancements to Windowing

By now, several improvements enhance windowing (Quinlan, 1993). Instead of using simple random samples at the beginning, the choice of initial windows is biased such that the distribution of classes is as close to uniform as possible. This strategy is similar to stratified simple random sampling and leads to more appropriate initial trees, if the distribution of classes in the original training set is unbalanced. Secondly, whereas first windowing versions fix the number of misclassified tuples which are added to the window in each cycle, the enhanced windowing mechanism uses at least half of all misclassified tuples. Thereby, windowing attempts to speed up convergence on a final tree. Thirdly, windowing is now able to stop iterative construction of decision trees before the resulting classifier assigns correct class labels to each tuple in the remaining test set, if the sequence of trees does not become more accurate. In noisy domains, early termination prevents an inexorable growth of the window until it includes almost all training tuples. Finally, Fürnkranz (1997) proposes more efficient windowing techniques for rule-learning algorithms rather than top down induction of decision trees and discusses problems of noisy data in windowing in more detail.

Advantages and Disadvantages

In terms of the unifying framework of focusing solutions, windowing applies (stratified) simple random sampling and uses wrapper evaluation. The specific evaluation criterion is the error rate on unseen tuples, and windowing repeats data mining on incrementally enlarged samples.

Quinlan points out that windowing has several advantages, even in situations where focusing is not necessary. In some domains, windowing leads to faster construction of decision trees. In other domains, windowing results in more accurate decision trees (Quinlan, 1993). However, systematic experiments by Wirth & Catlett (1988) show that using windowing is less efficient than top down induction of decision trees without windowing. Iterations on windows of increasing size often lead to larger overall numbers of tuples used during classifier generation than in a single run on the entire focusing input. Moreover, selection of misclassified tuples to enlarge the current window also contains an extra bias that possibly results in less representative samples in comparison to pure simple random sampling. Catlett (1991) confirms these results using the enhanced windowing mechanism.

4.1.2.5 Dynamic Sampling Using the PCE Criterion

Dynamic sampling using the *Probably Close Enough criterion* (PCE criterion) follows up windowing (John & Langley, 1996). In general, *dynamic sampling* covers all focusing solutions which use wrapper evaluation criteria and iteratively enhance the current focusing output until the evaluation criterion becomes true. In contrast, *static sampling* generates a focusing output only once and uses this focusing output for data mining. The PCE criterion is a specific evaluation criterion used in dynamic sampling. The idea in dynamic sampling using the PCE criterion is to select a focusing output that is probably good enough to yield the same quality of data mining results as if we use the entire focusing input.

Whereas windowing uses error rates of classifiers on remaining tuples in the focusing input for evaluation, the PCE criterion is weaker. The PCE criterion allows focusing to stop as soon as the probability that the difference between accuracies of classifiers generated on the focusing input and the current focusing output is higher than a first threshold no longer exceeds a second threshold. Similar to windowing, dynamic sampling using the PCE criterion iteratively enlarges the current focusing output until the PCE criterion holds. Note, the PCE criterion is only used for evaluation of the current focusing output. The generation of the focusing output again uses simple random sampling.

Tuple Selection in Dynamic Sampling Using the PCE Criterion

In terms of the framework for the definition of focusing tasks, the PCE criterion again refers to focusing on constrained subsets. The PCE criterion is true, if the probability that the difference between accuracies of classifiers generated on the focusing input and the current focusing output exceeds Δ_1 is not higher than Δ_2 (see definition 4.1.2). Thus, the two parameters Δ_1 and Δ_2 specify meanings of *close enough* and *probably*, respectively.[3]

Definition 4.1.2 (PCE Criterion)

Assume a (database) table (A, T), a focusing input $f_{in} \subseteq T$, a focusing output $f_{out} \subseteq f_{in}$, a data mining algorithm D, a probability function p on classification accuracies a, and $0 \leq \Delta_1, \Delta_2 \leq 1$. Predicate $P(f_{in}, f_{out})$ in dynamic sampling using the PCE criterion is

$$P(f_{in}, f_{out}) \quad :\Longleftrightarrow \quad p\left(a(D(f_{in})) - a(D(f_{out})) > \Delta_1\right) \leq \Delta_2.$$

---

[3]The PCE criterion is similar to the PAC learning paradigm (Valiant, 1984).

Advantages and Disadvantages

In terms of the unifying framework of focusing solutions, dynamic sampling using the PCE criterion utilizes simple random sampling and wrapper evaluation. Evaluation applies data mining algorithms to the current focusing output, estimates accuracies of resulting classifiers on an additional sample, and incrementally enlarges the focusing output, if the quality of the current focusing output is probably not good enough. Note, if data mining algorithms are not able to incrementally extend classifiers using additional tuples, dynamic sampling implies iterative data mining from scratch for each enlarged focusing output.

As a focusing solution, dynamic sampling using the PCE criterion is inefficient. Moreover, experimental results reported by John & Langley (1996) show only small improvements in terms of classification accuracies in comparison to static sampling while generating classifiers on larger samples than static sampling. The experimental procedure used to produce these results is also doubtful. In order to get large data sets, they *inflated* smaller data sets from the UCI repository by duplicating tuples 100 times. Although they argue that redundancy is a typical phenomenon in real-world applications, this strategy increases the probability that simple random samples, which contain more tuples than the original UCI database, include at least a single tuple within each set of duplicates. Then, the focusing output contains the same information as the original UCI database.

4.1.3   Clustering

The second step of the unifying framework uses *clustering* techniques to identify subsets of similar tuples within the focusing input. In statistics as well as in machine learning exist many efforts to identify and to describe groups of similar objects (see Bock (1974), Genari (1989), Genari *et al.* (1989), for example). The key idea to utilize clustering techniques for focusing purposes suggests to select only a few representatives for each cluster of similar objects. This set of representatives forms the focusing output, and we apply data mining algorithms to this focusing output. Note, if selection of representatives is part of clustering, these methods realize compound approaches which cover both clustering and prototyping in terms of the unifying framework. In the following, we briefly characterize principles of clustering techniques. We focus the description of clustering techniques on approaches in statistics (Bock, 1974; Bacher, 1994; Grimmer & Mucha, 1997).

Definition 4.1.3 (Clustering)

Assume a (database) table (A, T), a focusing input $f_{in} \subseteq T$, and L sets of tuples $C = \{c_1, c_2, \ldots, c_l, \ldots, c_L\}$.[4] C is a clustering of f_{in}, if

---

[4]Note, in this and all the following definitions, the focusing input might also be the result of a preliminary step.

$\forall c_l \in C: \quad (i) \quad c_l \neq \emptyset, \quad$ and $\quad (ii) \quad c_l \subseteq f_{in}.$

If C is a clustering of f_{in}, each $c_l \in C$ is a cluster in f_{in}.

Definition 4.1.3 specifies clusterings as sets of non-empty subsets of the focusing input. Note, this definition also allows single sets with single tuples as clusterings. Subsequently, we refine this basic definition and add more conditions that make clusterings more reasonable.

Definition 4.1.4 (Cluster Homogeneity and Heterogeneity)

Assume a (database) table (A, T), a focusing input $f_{in} \subseteq T$, and two clusters $c_l, c_{l'}$ in f_{in}, $c_l \neq c_{l'}$. We define homogeneity within cluster c_l and heterogeneity between clusters c_l and $c_{l'}$ as two alternative functions

$Hom_1 : T^{\subseteq} \longrightarrow [0; 1],$

$$Hom_1(c_l) := \frac{1}{(\mid c_l \mid)^2} \cdot \sum_{i=1}^{\mid c_l \mid} \sum_{i'=1}^{\mid c_l \mid} Sim(t_i, t_{i'}),$$

$Hom_2 : T^{\subseteq} \longrightarrow [0; 1],$

$$Hom_2(c_l) := \min_{i,i'=1}^{\mid c_l \mid} \{Sim(t_i, t_{i'})\},$$

$Het_1 : T^{\subseteq} \times T^{\subseteq} \longrightarrow [0; 1],$

$$Het_1(c_l, c_{l'}) := \frac{1}{\mid c_l \mid \cdot \mid c_{l'} \mid} \cdot \sum_{i=1}^{\mid c_l \mid} \sum_{i'=1}^{\mid c_{l'} \mid} Sim(t_i, t_{i'}),$$

$Het_2 : T^{\subseteq} \times T^{\subseteq} \longrightarrow [0; 1],$

$$Het_2(c_l, c_{l'}) := \max_{i=1, i'=1}^{\mid c_l \mid, \mid c_{l'} \mid} \{Sim(t_i, t_{i'})\}.$$

In order to measure the quality of clusterings, we first consider similarities within each cluster and similarities between pairs of clusters. Intuitively, we are interested in clusterings that group sufficiently similar tuples into clusters and separate adequately dissimilar tuples into different clusters. Thus, similarities within each cluster should be higher than similarities between pairs of clusters. We define *homogeneity* within clusters and *heterogeneity* between clusters to specify this intuition precisely. Homogeneity and heterogeneity allow evaluation of the quality of clustering structures.

Definition 4.1.4 states two alternative functions to measure homogeneity within clusters and heterogeneity between clusters, respectively. The first alternatives, Hom_1 and Het_1, use average similarities between tuples within a cluster to measure homogeneity and average similarities between tuples of different clusters to measure heterogeneity. Note, these measures weigh all occurring similarities equally, and effects of especially high or especially low similarities on homogeneity and heterogeneity eliminate each other. In contrast, the second alternatives use minimum similarities between tuples within a cluster to measure homogeneity and maximum similarities between tuples of different clusters to measure heterogeneity, respectively. These measures consider the worst occurring similarities according to homogeneity and heterogeneity. Hence, Hom_2 and Het_2 are stronger than Hom_1 and Het_1, and they require higher quality of clustering structures.

4.1.3.1 Clustering Algorithms

In general, clustering algorithms generate clusterings. In this section, we discuss several criteria to distinguish different clustering algorithms in statistics according to characteristics of their outputs. Figure 4.2 outlines these criteria in order to classify clustering algorithms according to the following aspects:

- *Coverage*

  Exhaustive clustering algorithms generate clusterings such that each tuple in the focusing input is at least member of a single cluster, whereas *non-exhaustive* clustering algorithms build clusterings which do not cover each tuple.

- *Separation*

  Overlapping clustering algorithms produce clusterings such that at least a single tuple belongs to more than a single cluster, whereas *non-overlapping* clustering algorithms separate the focusing input into disjoint groups of similar tuples.

- *Structure*

  Hierarchical clustering algorithms create clusterings at different levels of granularity, whereas *non-hierarchical* clustering algorithms only return a single clustering at one specific level. In case of hierarchical clustering algorithms, we also distinguish whether we build clustering hierarchies bottom-up or top-down. *Agglomerative* approaches start with a single cluster for each tuple in the focusing input and iteratively combine clusters, whereas *divisive* algorithms begin with a single cluster containing all tuples and continue with iterative separations of clusters.

The following definitions describe these criteria in more detail. Each definition precisely states characteristics of specific types of clustering algorithms, and we discuss the appropriateness of each aspect in the context of focusing.

Definition 4.1.5 indicates the notion of exhaustive clusterings. Clusterings of the focusing input are exhaustive, if for each tuple in the focusing input exists a

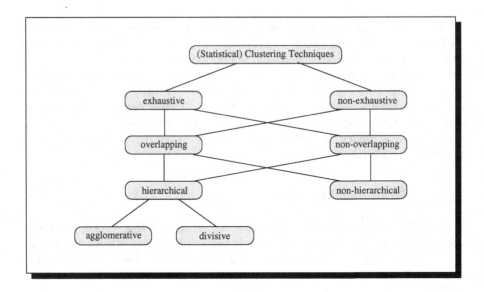

Figure 4.2: Classification Criteria for Clustering Techniques in Statistics

cluster such that the tuple is an element of this cluster. Otherwise, clusterings are non-exhaustive. For focusing purposes and the selected subset of focusing tasks, exhaustive clusterings are preferable since we attempt to represent the information in the focusing input completely. If we utilize clustering in focusing, and the resulting clustering does not represent each tuple, information of non-represented tuples is not part of the focusing output unless their information is redundant.

Definition 4.1.5 (Exhaustive Clustering)

Assume a (database) table (A, T), *a focusing input* $f_{in} \subseteq T$, *and a clustering* $C = \{c_1, c_2, \ldots, c_l, \ldots, c_L\}$ *of* f_{in}. *C is an exhaustive clustering of* f_{in}, *if*

$$\forall t_i \in f_{in} \; \exists c_l \in C : \; t_i \in c_l.$$

Definition 4.1.6 states additional characteristics of non-overlapping clusterings. Clusterings of the focusing input are non-overlapping, if the intersection of each pair of distinct clusters is empty. Otherwise, clusterings are overlapping. In the context of focusing, we prefer non-overlapping clusterings since the representation of information within clusters by selecting only a few prototypes is more difficult, if clusters and hence their information overlap.

Definition 4.1.6 (Non-Overlapping Clustering)

Assume a (database) table (A, T), *a focusing input* $f_{in} \subseteq T$, *and a clustering* $C = \{c_1, c_2, \dots, c_l, \dots, c_L\}$ *of* f_{in}. C *is a non-overlapping clustering of* f_{in}, *if*

$$\forall c_l, c_{l'} \in C, \ l \neq l' : \ c_l \cap c_{l'} = \emptyset.$$

Finally, definition 4.1.7 specifies the concept of hierarchical clusterings. Clusterings of the focusing input are hierarchical, if for each pair of distinct clusters their intersection is either empty or one cluster is a subset of the other cluster. Otherwise, clusterings are non-hierarchical. The main advantage of hierarchical clusterings for focusing purposes is the hierarchical structure of clusters. This structure defines several clusterings at different levels of granularity. For representations of clusters, focusing approaches are then able to select the most appropriate level in the hierarchy. This ability improves the quality of representativeness since it is possible to refine particular clusters, if it is not possible to represent their information appropriately. On the other hand, computational efforts to generate hierarchical clusterings are more expensive than for non-hierarchical clusterings. This aspect makes hierarchical clustering approaches less feasible for focusing solutions.

Definition 4.1.7 (Hierarchical Clustering)

Assume a (database) table (A, T), *a focusing input* $f_{in} \subseteq T$, *and a clustering* $C = \{c_1, c_2, \dots, c_l, \dots, c_L\}$ *of* f_{in}. C *is a hierarchical clustering of* f_{in}, *if*

$$\forall c_l, c_{l'} \in C, \ l \neq l' : \ c_l \cap c_{l'} = \emptyset \ \lor \ c_l \subseteq c_{l'} \ \lor \ c_{l'} \subseteq c_l.$$

In summary, we regard exhaustive, non-overlapping, hierarchical clustering algorithms as most appropriate for focusing purposes. However, since we are interested in efficient focusing solutions and hierarchical clustering algorithms are more expensive in terms of running time, we propose to use exhaustive, non-overlapping, non-hierarchical clustering algorithms. In the following two subsections, we exemplify a particular exhaustive, non-overlapping, non-hierarchical clustering technique and present an example for utilization of clustering as part of an existing focusing solution in KDD. For the development of more intelligent sampling techniques, we follow up positive aspects of both approaches.

4.1.3.2 Leader Clustering

Leader clustering constructs exhaustive, non-overlapping, non-hierarchical clusterings of the focusing input (Hartigan, 1975). Therefore, we assume a similarity measure to compute similarities between pairs of tuples, and a similarity threshold δ. The algorithm generates a partition of all tuples into clusters and selects a leader tuple for each cluster, such that each tuple in a cluster is within a distance of $1 - \delta$ from the leader tuple. Thus, threshold δ is a measure of the radius of each cluster, and leader clustering is a specific compound approach to clustering and prototyping, if we consider each leader tuple as the prototype of its cluster.

Algorithm 4.1.4   LEACLU(f_{in}, Sim, δ)

```
begin
    c₁ := {t₁};
    l₁ := t₁;
    L := 1;
    i := 1;
    while i ≤ | f_in | do
            i := i + 1;
            i := 1;
            while i ≤ L do
                    if Sim(l_i', t_i) ≥ δ
                        then c_i' := c_i' ∪ {t_i};
                            break;
                        else i := i + 1;
                    endif
            enddo
            if i > L
                then c_i' := {t_i};
                    l_i' := t_i;
                    L := L + 1;
            endif
    enddo
    return({c₁, ..., c_l, ..., c_L});
end;
```

Algorithm 4.1.4 shows an implementation of leader clustering. In general, leader clustering makes one pass through the focusing input and assigns each tuple to the cluster of the first leader tuple which is sufficiently similar to this tuple. It creates new clusters and new leader tuples for tuples which are not close enough to any existing leader tuple. This particular implementation of leader clustering works as follows. The first cluster initially contains the first tuple in the focusing input, and its leader is also the first tuple. While leader clustering has not seen all tuples in the focusing input, it checks similarities between the next tuple in the focusing input and all existing leaders. As soon as leader clustering observes the first leader with similarity higher than δ to the next tuple, it assigns the current tuple to the respective cluster of this leader tuple. If no leader tuple exists which exceeds similarity threshold δ to the next tuple, LEACLU creates a new cluster with this tuple as its leader tuple.

Advantages and Disadvantages

In terms of the unifying framework of focusing solutions, leader clustering is an implementation of the clustering step and overlaps with prototyping. Leader clustering builds clusters and leader tuples at the same time. Leader tuples correspond to prototypes, and each leader represents a single cluster. Thus, the set of all leaders defines a set of prototypes that represents the entire focusing input. Note, leader clustering does not apply any evaluation criterion.

The positive feature of leader clustering is that it only requires one sequential pass through the data. Thus, it is not necessary to keep the whole focusing input in main memory. Leader clustering only needs to retain leader tuples and an index structure to hold resulting clusters. On the other hand, several negative properties follow from immediate assignments of tuples to clusters as soon as the first leader exceeds similarity threshold δ.

The first negative property is that clusterings are not invariant with respect to reordering of tuples. For example, the first tuple is always the leader tuple of the first cluster. The second negative property is that the first clusters are always likely to contain more tuples than later clusters, since they get first chance at each tuple as it is allocated. Thirdly, representativeness of leaders mainly depends on similarity threshold δ. For small values of δ, leader clustering selects only a few leaders, whereas for large values of δ, the set of leaders is likely to contain many tuples. Increasing δ usually results in an increasing number of leaders, and it is difficult to set δ to the most appropriate value in terms of representativeness of resulting leaders.

4.1.3.3 CLARANS

CLARANS (Clustering Large Applications based on RANdomized Search) is the only approach in the state of the art towards focusing solutions that explicitly refers to focusing in KDD (Ng & Han, 1994; Ester *et al.*, 1995). CLARANS is a specific approach to clustering which follows up the idea of *k-means* clustering techniques in statistics.

In k-means clustering, the initial clustering of the focusing input is a random selection of k tuples and assignments for each remaining tuple to its closest representative in the initial set of k tuples. In the next step, k-means clustering computes mean tuples within each cluster. Means of all attribute values within a cluster for each attribute constitute a mean tuple. Thus, this strategy is only applicable in case of quantitative attributes. The set of means is then the set of representatives of clusters in the next iteration. Now, k-means clustering assigns each tuple to its closest mean tuple again and compares the quality of the resulting clustering to the previous structure. If the clustering quality increases, the current clustering is replaced by the new clustering. Usually, sums of squared distances within clusters are the quality criterion for this comparison. This process iterates until no more quality improvement is possible.

The main drawback of k-means clustering is its dependency on concepts of means. Since means only exist for quantitative attributes, this clustering technique is not applicable to domains with qualitative attributes. Moreover, mean tuples as representatives of clusters refer to artificial tuples, if the means of attribute values for each attribute do not correspond to an existing tuple in the (database) table. Enhancements to k-means clustering overcome these disadvantages by using *medoids* instead of mean tuples to represent clusters. The definition of medoids only relies on a notion of distance (or similarity, respectively). In a set of initial tuples, the medoid of an arbitrary tuple is the tuple with minimum distance to the currently considered tuple. The clustering replacement step in k-medoid clustering, which corresponds to computation of means in k-means clustering, is a random replacement of a medoid by a non-medoid tuple. CLARANS uses such a medoid clustering approach with euclidian distance.

Algorithm 4.1.5   CLARANS($f_{in}, m, loops, neighbors$)

---

begin
  $loop := 1$;
  $C\ := \emptyset$;
  $quality := max\_quality$;
  while $loop \leq loops$ **do**
      $C := initialize(f_{in}, m)$;
      $neighbor := 1$;
      while $neighbor \leq neighbors$ **do**
          $C\ := exchange(C, f_{in})$;
          if $quality(C\ ) - quality(C) < 0$
            then $C := C\ $;
                $neighbor := 1$;
             else $neighbor := neighbor + 1$;
          endif
      enddo
      if $quality(C) < quality$
          then $quality := quality(C)$;
              $C\ := C$;
      endif
      $loop := loop + 1$;
  enddo
  return($C\ $);
end;

Algorithm 4.1.5 shows an implementation of CLARANS. After initialization of the number of loops, the current best clustering, and the current best clustering quality, CLARANS randomly generates an initial clustering within each loop.[5] Therefore, procedure *initialize* randomly selects m tuples in the focusing input

---

[5]Note, the quality measure in CLARANS yields low values, if the quality of the

and assigns each of the remaining tuples to its closest tuple in this random selection using euclidian distance between tuples. Hence, the random sample of m tuples initiates the representative set of medoids and assignments of remaining tuples create the initial clustering structure.

Within the inner loop, CLARANS iteratively checks at most *neighbors* alternative neighbor clusterings according to quality improvements of the current clustering. Procedure *exchange* generates a neighbor clustering of C by replacement of a randomly selected medoid in C with a randomly selected tuple in the set of non-medoids and an update of assignments of remaining tuples to their closest medoids. If the quality of the resulting clustering is better than the quality of the current clustering, CLARANS replaces the old clustering with the new clustering. In this approach, clustering quality is the sum of squared distances between medoids and members in their clusters. Thus, we search for minimum values of this sum, and quality improvement occurs, if the difference between sums is negative.

CLARANS iterates local optimization in the neighborhood of the current clustering until it is not able to improve quality *neighbors* number of times. At the end of local search, CLARANS compares the local optimum with the current overall best clustering structure C^*. If the local optimum is more appropriate than the current best clustering, CLARANS updates the current overall best clustering structure. Local search and potential replacements of the current overall best clustering structure continues until a maximum number *loops* of local searches is complete.

In summary, CLARANS performs hill-climbing in a space of clusterings with random generation of new points in the neighborhood of the current point in the search space and random jumps after each local optimization.

Advantages and Disadvantages

In terms of the unifying framework of existing focusing solutions, CLARANS includes a clustering step with local evaluation in terms of clustering quality. Since random selections guide the search in the space of all possible clusterings, CLARANS inherently incorporates a sampling step as well. If we consider the final set of medoids as representative prototypes, this approach also realizes a prototyping step with (random) prototype selection. All in all, CLARANS is an example instantiation of the unifying framework that combines all three steps to a compound focusing solution.

Although iterative optimization of the current best clustering structure is an advantage in CLARANS, the fixed number of clusters and prototypes is one of its main drawbacks. An additional disadvantage is random generation of initial clusterings and random replacements of medoids to improve the current

---

clustering is high. Hence, the maximum possible quality value represents the worst clustering.

clustering structure. If the number of loops and the number of neighbors are small, CLARANS fails to find an optimum in the space of all possible clusterings with high probability. If CLARANS uses higher values for the number of loops and the number of neighbors, it is likely to generate more appropriate clusterings, but CLARANS becomes inefficient in terms of running time at the same time.[6]

4.1.4 Prototyping

The third step in the unifying framework of focusing solutions is *prototyping*. This step refers to selection or construction of prototypes. In general, prototypes are more condensed descriptions of sets of tuples. Prototyping assumes that a single tuple is able to represent information of an entire subset of tuples. Note, prototypes are not necessarily the most typical or most frequent tuples among tuples (Tversky, 1977).

4.1.4.1 Prototype Selection and Prototype Construction

In prototyping, we distinguish between the following types of approaches:

- *Prototype Selection*

  Prototype selection results in a subset of the original set of tuples and is often based on notions of *prototypicality*. For each tuple in the original subset of tuples, prototypicality measures the degree of prototypicality of this tuple among all tuples in the subset (Zhang, 1992). Then, we select those tuples which maximize prototypicality as the representative subset of prototypes (see Skalak (1993), (1994), Smyth & Keane (1995) for further examples).

- *Prototype Construction*

  Prototype construction generates prototypes which do not necessarily refer to existing tuples in a (database) table. Therefore, this type of prototyping uses a specific function to explicitly build new tuples that represent information of an entire subset of tuples. For example, if all attributes are quantitative, we average all occurring values for each attribute separately, and the sequence of average values constitutes the center of all tuples as the prototypical representation (see Linde *et al.* (1980), Barreis (1989), Sen & Knight (1995), Börner *et al.* (1996) for further examples).

In terms of the framework for the definition of focusing tasks, prototype selection as well as prototype construction define specific predicates or functions

---

[6]Note, Ester *et al.* (1995) define enhancements to the basic CLARANS algorithm in order to overcome these efficiency problems. However, since these enhancements mostly concern technical issues in database technology, we do not discuss these extensions in detail.

that refer to the respective predicate or function in the definition of focusing outputs (see definition 3.2.3, page 50). Prototype selection criteria correspond to predicates of constrained subsets, whereas prototype construction requires definitions of functions that specify construction principles for constructed sets.

As the enumeration of example approaches to prototyping indicates, many approaches exist for both, prototype selection and prototype construction. Subsequently, we outline two specific example approaches which particularly contribute to the development of more intelligent sampling techniques in more detail.

4.1.4.2 Instance-Based Learning

We already mentioned extensions of nearest neighbor classifiers that attempt to reduce storage requirements and to make retrieval of nearest neighbors more efficient when we described basic nearest neighbor classifiers (see section 2.5.3, page 37). These extensions either select a smaller subset of the entire training set, or they construct artificial representations of subsets of tuples in the case base. One of the first attempts to follow the second alternative is *edited kNN* (Chang, 1974; Dasarathy, 1991). Edited kNN constructs prototypes by merging tuples to their mean representations. However, this approach results in artificial prototypes and is only applicable in domains which only contain quantitative attributes. Consequently, we do not consider edited kNN in more detail, since it does not contribute to solutions for the selected focusing tasks.

Algorithm 4.1.6   IBL2(f_{in}, Sim)

---

begin
  $f_{out} := \{t_1\}$;
  $i := 1$;
  while $i \leq \mid f_{in} \mid$ **do**
      $i := i + 1$;
      if $NN(f_{out}, t_i, Sim) \neq t_{iN}$
        then $f_{out} := f_{out} \cup \{t_i\}$;
      endif
  enddo
  return(f_{out});
end;

Instance-based learning (IBL) algorithms follow the first approach (Aha *et al.*, 1991). The family of IBL algorithms starts with IBL1 which is identical to simple nearest neighbor classifiers described in section 2.5.3. IBL2 is an extension of IBL1 and reduces storage requirements. IBL3 is another extension and tries to cope with problems of noisy tuples. Note, the original purpose of instance-based learning is classification not focusing. However, IBL2 essentially implements a focusing solution, if we consider retained tuples as the focusing output. Since we are interested in focusing related issues, we only consider IBL2 here.[7]

Algorithm 4.1.6 presents an implementation of IBL2. After initialization of the focusing output, IBL2 iteratively considers each tuple in the focusing input. For each tuple, IBL2 retrieves the class label of the most similar tuple in the current focusing output. If this class label is the same as the class label of the currently considered tuple, i.e., if the current focusing output is already able to classify this tuple correctly, IBL2 proceeds with the next tuple. If both class labels do not coincide, the new tuple is added to the focusing output. IBL2 is identical to IBL1 except that it only retains misclassified tuples.

Prototype Selection in Instance-Based Learning

In terms of the framework for the definition of focusing tasks, IBL2 specifies a concrete criterion for selection of constrained subsets as focusing outputs. Definition 4.1.8 states this specific prototype selection criterion in IBL2 more precisely. Note, in this definition, we consider the set of retained tuples as the final focusing output. The prototype selection criterion in IBL2 evaluates to true, if for all tuples in the focusing input, a tuple in the focusing output exists with the same class label, and for all pairs of distinct tuples in the focusing output, class labels do not coincide, if one tuple is the nearest neighbor of the other tuple. This means that each tuple in the focusing output is necessary to classify all tuples correctly.

Definition 4.1.8 (Prototype Selection in IBL2)

Assume a (database) table (A, T), a class attribute $a_N \in A$, a focusing input $f_{in} \subseteq T$, and a focusing output $f_{out} \subseteq f_{in}$. Predicate $P(f_{in}, f_{out})$ in IBL2 is

$$P(f_{in}, f_{out}) \quad :\Longleftrightarrow$$

$$\forall t_i \in f_{in} \ \exists t_{i'} \in f_{out} : \quad t_{iN} \equiv t_{i'N}$$

$$\wedge \qquad \forall t_i, t_{i'} \in f_{out} : \quad t_i \neq t_{i'} \ \wedge \ t_i = NN(f_{out}, t_{i'}, Sim)$$

$$\Longrightarrow \ t_{iN} \neq t_{i'N}.$$

Advantages and Disadvantages

In terms of the unifying framework of focusing solutions, IBL2 uses prototyping in form of a prototype selection approach. It neither utilizes sampling techniques, nor does it apply clustering algorithms. Prototype selection in IBL2 is similar to leader selection in leader clustering. IBL2 selects tuples as pro-

---

[7]In case-based reasoning, pre-selection strategies also try to reduce the number of tuples before retrieval of most similar cases in the case base. We refer to Goos (1995) for a survey of pre-selection strategies in case-based reasoning.

totypes, if their nearest neighbors in the current focusing output have different class labels, whereas leader clustering retains tuples as new leaders, if similarities between all leaders and the currently considered tuple are smaller than similarity threshold δ. IBL2 does not use an explicit evaluation criterion. However, the selection criterion inherently evaluates the current focusing output and iteratively enlarges the focusing output, if its current quality is not sufficient yet, i.e., if the current focusing output is not able to classify all remaining tuples correctly. In this sense, IBL2 is similar to windowing and dynamic sampling using the PCE criterion.

The main advantage of IBL2 is its straightforward selection of tuples. IBL2 only requires one sequential pass through the data, and most of the complexity in IBL2 is due to computations of similarities. On the other hand, there is no guarantee for a small number of tuples in the focusing output. Moreover, IBL2 considers information of all attributes to retrieve the most similar tuple in the current focusing output, but only uses information on the class attribute to decide whether to retain tuples or not. This approach is probably appropriate for nearest neighbor classifiers, but for different data mining algorithms, focusing solutions benefit from consideration of additional information of other attributes when deciding whether to retain tuples or not.

4.1.4.3 PL

The second example of prototyping is PL (Datta & Kibler, 1995; Datta, 1997). This approach learns prototypical concept descriptions for sets of tuples. Although the final output of PL contains prototypical concept descriptions for each class, PL uses a top-down splitting mechanism to partition tuples for each class separately, and then learns prototypes for each partition. The disjunction of all prototypes for each class forms the prototypical concept description of this class. Note, the original purpose of PL is classification, not focusing. Hence, generation of prototypes in PL corresponds to generation of classifiers.

Algorithm 4.1.7 shows an implementation of PL. After initialization of the focusing output to an empty set, PL generates prototypes for each class separately. First, PL partitions all tuples in the focusing input which belong to the same class with the current index k into subsets $S := \{s_1, \ldots, s_l, \ldots, s_L\}$. Then, PL constructs a single prototype for each of the resulting subsets. At the end, PL returns the set of all prototypes as the focusing output.

Similar to top down induction of decision trees, the *partition* step recursively separates sets of tuples into subsets according to attribute values. For qualitative attributes, the partition contains subsets of tuples which all have the same value within each subset. For quantitative attributes, the partition consists of subsets of tuples which all have either values below a selected threshold or values above this threshold within each subset. The selection of the next partition attribute corresponds to the selection of the test attribute in top down induction of decision trees. However, the evaluation measure to select the *best* attribute is based on a quality measure for partitions which is different from information gain in top down induction of decision trees. The idea of this quality measure is to partition tuples into subsets, if the randomness of attribute values within each set of tuples decreases. Therefore, PL uses the binomial distribution to estimate the probability that partitioning decreases randomness of attribute values for attributes which are not used as the partition attribute. The attribute which results in the partition of highest quality is the *best* attribute. Recursive partition iterates until no more quality improvement is possible.[8]

Algorithm 4.1.7 $PL(f_{in}, partition, prototype)$

---

begin
   $f_{out} := \emptyset$;
   $k := 1$;
   while $k \leq N_N$ **do**
        $S := partition(\{t_i \in f_{in} \mid t_{iN} \equiv a_{Nk}\})$;
        $l := 1$;
        while $l \leq \mid S \mid$ **do**
            $f_{out} := f_{out} \cup prototype(s_l)$;
            $l := l + 1$;
        enddo
        $k := k + 1$;
   enddo
   return(f_{out});
end;

Prototype Construction in PL

In terms of the framework for the definition of focusing tasks, PL corresponds to focusing on constructed sets. Note, although PL contributes to solutions for focusing tasks with constructed sets as focusing outputs rather than to solutions for the selected focusing tasks, we consider parts of PL as a promising state of the art approach which is relevant for the development of focusing solutions for simple subsets and constrained subsets as well.

Definition 4.1.9 (Prototype Construction in PL)

Assume a (database) table (A, T), *a focusing input* $f_{in} \subseteq T$, *a clustering* $S = \{s_1, \ldots, s_l, \ldots, s_L\}$ *of* f_{in}, *and a focusing output* $f_{out} \subseteq f_{in}$. *Function* $P(f_{in}, f_{out})$ *in PL is*

$$P(f_{in}, f_{out}) \quad :\Longleftrightarrow \quad \forall t_i \in f_{out} \; \exists s_l \in S : \; t_i \equiv prototype(s_l)$$

---

[8]An extension of PL, called *Symbolic Nearest Mean*, uses k-means clustering to generate partitions of tuples before prototyping (Datta & Kibler, 1997).

$$\text{with} \quad prototype(s_l) \quad := \quad (t_{i^*1}, \ldots, t_{i^*j}, \ldots, t_{i^*N}),$$

$$\text{and} \quad t_{i^*j} \quad := \quad \begin{cases} m_j(s_l), & \text{if } a_j \text{ qualitative} \\ \mu_j(s_l), & \text{else} \end{cases}$$

Definition 4.1.9 describes prototype construction in PL. For each predictive attribute, PL computes a single cumulative value of all tuples within each subset of the partition. For qualitative attributes, this cumulated value is the modus, i.e., the value that most often occurs in the subset. For quantitative attributes, the cumulated value is the mean of all values. Since PL generates prototypes for each class separately, the class label of each prototype corresponds to the class label within its subset of tuples. Note, in addition to cumulative values for each attribute, each prototype includes strength values for each attribute. Strength values correspond to randomness of attribute value distributions within represented subsets of tuples. Low strength indicates high randomness. Attributes with strengths below a threshold are not part of prototypical concept descriptions. In this sense, definition 4.1.9 simplifies prototype construction in PL, since it does not include strength values.

Advantages and Disadvantages

In terms of the unifying framework of focusing solutions, PL performs clustering in combination with prototyping. The partition step in PL constructs exhaustive, non-overlapping, hierarchical clusters, and the prototyping step constructs a single artificial prototype for each cluster. Except the inherent evaluation within the partition step, PL does not apply any evaluation criterion.

PL is similar to edited nearest neighbor approaches. However, its advantage is the straightforward strategy to identify subsets of sufficiently similar tuples which are then represented by single prototypes. The main disadvantage of PL is the complexity of the partition step which is mainly due to expensive computations for quality assessments of partitions. Moreover, PL's focusing output does not necessarily refer to existing tuples in the original data set.

4.2 More Intelligent Sampling Techniques

The analysis of the state of the art of most promising efforts towards focusing solutions and the resulting unifying framework show interesting ideas but also incorporate a set of disadvantages. In order to develop more intelligent sampling techniques, we propose to follow up the unifying framework of focusing solutions. Intelligent sampling techniques cover positive aspects of state of the art efforts, whereas they overcome their negative properties. In this section, we develop more intelligent sampling techniques as enhancements to the unifying view of existing approaches towards focusing solutions.

Table 4.1: Existing Reusable Components in State of the Art Efforts

Step	Name	Origin	Purpose	Advantage	Disadvantage	Reuse	Reference
Sampling		statistics	sampling surveys	efficient	uninformed, non-deterministic	random and systematic sampling, stratification	Cochran, 1977
	windowing	machine learning	sampling, classification	context sensitive	inefficient	none	Quinlan, 1993
	PCE criterion	KDD	sampling, classification	context sensitive	inefficient	none	John & Langley, 1996
Clustering		statistics	segmentation	grouping of similar tuples	partly inefficient	leader clustering	Bock, 1974
	leader clustering	statistics	segmentation	efficient	order-sensitive, fixed threshold	in improved form	Hartigan, 1975
	CLARANS	KDD	segmentation, focusing	subset of search space	fixed number of clusters and prototypes, random search	subset of search space	Ester et al., 1995
Prototyping		statistics, machine learning	prototype selection or construction	condensed representation	partly artificial prototypes	prototype selection	
	IBL2	machine learning	prototype selection, classification	efficient, reduced storage requirements	limited consideration of available information	prototype selection, similarities	Aha et al., 1991
	PL	machine learning	prototype construction, classification	partition	complexity of partition, artificial prototypes	partition idea	Datta & Kibler, 1995

In comparison to existing efforts towards focusing solutions, these enhancements are more intelligent since they take into account more information in the focusing input as well as parts of the focusing context. In this sense, less intelligent approaches do not sufficiently take into account these aspects. Consequently, existing approaches are less appropriate as solutions for the selected focusing tasks.

4.2.1 Existing Reusable Components

In summary, we now review the main advantages and disadvantages of state of the art efforts towards focusing solutions and outline components that we reuse for the development of more intelligent sampling techniques. Table 4.1 summarizes the main characteristics of each approach described previously and its relation to the unifying framework. For each approach, this table depicts its major step in the unifying framework, its name, its origin, its primary purpose, its main positive and negative features, its reusable aspect, and the respective reference in the literature. We stress the following properties of existing efforts towards focusing solutions:

- *Sampling*

    - The focus of statistical sampling techniques is supporting sampling surveys in human populations. Sampling techniques are efficient in terms of running time but they are uninformed since they do not take into account the focusing context, and they implement non-deterministic focusing solutions. We reuse the idea of a sampling step and provide two statistical sampling techniques, random and systematic sampling, as part of the enhancements to existing focusing solutions. Sampling techniques either help to get an initial overview on the focusing input or work as part of multi-step solutions in order to select intermediate working sets. We also reuse the idea of stratification and propose a particular procedure to stratify tuples.

    - The origin of windowing and dynamic sampling using the PCE criterion is machine learning and KDD, respectively. Both approaches implement simple random sampling with wrapper evaluation for classification tasks. They iteratively apply data mining algorithms to the current focusing output and enlarge the current focusing output, if the resulting classifier does not meet the evaluation criterion. Although these approaches take into account the focusing context, they are less efficient than their basic sampling technique since they incorporate data mining as part of their focusing solution. Hence, both approaches do not solely prepare data before data mining. Although it is important to consider the focusing context before the development or selection of appropriate focusing solutions, we recommend not to integrate data mining algorithms into focusing solutions for efficiency reasons. Consequently, we do not follow up wrapper approaches.

- *Clustering*

    - Statistical clustering techniques help to identify subsets of sufficiently similar tuples in the focusing input in order to solve segmentation goals. If we select a small set of representatives within each cluster and unify these prototypes to constitute the final focusing output, we assume that the entire information encapsulated in subsets of sufficiently similar tuples is not completely necessary to generate appropriate data mining results. Instead, this approach presumes that subsets of representatives contain enough information to reveal suitable data mining results. Although many clustering techniques are computationally expensive, we partly reuse a specific clustering approach.

    - We propose to utilize leader clustering as part of the enhancements to existing focusing solutions since leader clustering is an efficient implementation among statistical clustering algorithms. In order to overcome its primary disadvantages, order-sensitivity and fixed similarity thresholds, we propose an improved variant of leader clustering.

    - CLARANS is the only existing approach in KDD which explicitly refers to focusing. We reuse its idea of working on subsets of the entire search space of focusing outputs as CLARANS' most positive feature, but we overcome its limitation on fixed numbers of clusters and prototypes. We propose to utilize systematic samples with increasing start positions as intermediate working sets within more intelligent sampling. Thereby, we ensure that working sets are spread across the entire focusing input and systematically cover the entire search space of focusing outputs.

- *Prototyping*

    - Prototyping approaches in statistics and machine learning provide mechanisms for prototype selection or prototype construction. Their general advantage is the generation of more condensed descriptions of the entire focusing input. Since we envisage focusing outputs that still contain original entities of the focusing input, we propose to prefer prototype selection instead of prototype construction.

    - In instance-based learning, prototype selection takes into account similarities to retrieve the most similar tuple within the current focusing output and to reduce storage requirements of the final focusing output by storing only instances which are necessary to classify all tuples correctly. However, IBL2 only uses information on class labels to decide whether new candidates in the focusing input become members of the focusing output, although similarities consider all attributes.

We follow up the idea of prototype selection and using similarities, but we consider information of all attributes when deciding whether to retain tuples or not rather than only the class attribute.

– Although PL implements prototype construction for classification purposes and we prefer to utilize prototype selection, we also reuse the general idea to partition the focusing input before prototyping. We realize a stratification procedure which generates comparable partitions of the focusing input but is less complex than partition in PL.

In summary, the key idea of more intelligent sampling assumes the existence of subsets of sufficiently similar tuples within the focusing input. We presume that each subset contains more information than necessary to generate appropriate data mining results, and that smaller sets of prototypes are able to represent the entire information in the focusing input. Hence, we replace each subset of sufficiently similar tuples by representative prototypes, and we use these prototypes as input to the subsequent data mining phase.

The following subsections describe enhancements to the unifying framework of focusing solutions and develop specific approaches which implement more intelligent sampling techniques. *Advanced leader sampling* (section 4.2.2) follows up the idea of leader clustering (Reinartz, 1997). Two enhancements in advanced leader sampling to prepare the focusing input before focusing overcome the main deficits of leader clustering. *Similarity-driven sampling* (section 4.2.3) contains additional enhancements along these lines that automatically estimate and adapt similarity thresholds in leader clustering and allow elimination of less representative prototypes as well as weighting of remaining prototypes in the focusing output (Reinartz, 1998). All in all, enhancements to the unifying framework lead to an extended unified approach to focusing solutions.

4.2.2 Advanced Leader Sampling

In order to implement the key idea of more intelligent sampling techniques, we first apply clustering to identify subsets of sufficiently similar tuples. Since we normally deal with large data sets, we reuse a specific clustering algorithm that is both appropriate for our purpose and sufficiently efficient. Leader clustering is a good candidate since it clusters data in a single pass. Thus, we use leader clustering as a starting point.

A straightforward approach to follow up leader clustering as a focusing solution is the selection of all leaders in constructed clusters as the final focusing output. The resulting strategy, called *leader sampling*, suffers from the following drawbacks:

• Leader sampling is sensitive to the order of tuples within the focusing input. Tuples with small indices have higher probabilities to get into the final focusing output.

- Leader sampling is less efficient, if the set of leaders becomes large. Since leader sampling compares each candidate to all leaders in the worst case, this comparison possibly costs many similarity computations.

- Leader sampling largely depends on similarity threshold δ. If we set this threshold to an inappropriate value, the resulting focusing output does not represent the focusing input well.

In the following, we propose two mechanisms, *sorting* and *stratification*, in order to overcome the first two drawbacks.[9] We first outline the resulting *advanced leader sampling* approach which uses the straightforward selection strategy of leader sampling but incorporates sorting and stratification to prepare the focusing input in advance. Then, we describe each component in detail.

4.2.2.1 Algorithm ALESAM

Algorithm 4.2.1 describes an implementation of advanced leader sampling. After initialization of the focusing output to an empty set, this focusing solution sorts the focusing input according to order relation $\succ$ (see section 4.2.2.3) in order to eliminate the strong order-sensitivity of leader sampling. In the next step, ALESAM stratifies the focusing input into subsets of similar tuples (see section 4.2.2.4). Leader sampling is then iteratively applied to each resulting stratum s_l separately. The unification of resulting leader samples within each stratum constitutes the final focusing output.

Algorithm 4.2.1   ALESAM($f_{in}, Sim, \delta, max, discretize$)

begin
  $f_{out} := \emptyset$;
  $f_{in} := \text{SORT}(f_{in}, \succ)$;
  $S := \text{STRATIFY}(f_{in}, N, max, discretize)$;
  $l := 1$;
  while $l \leq |S|$ **do**
        $f_{out} := f_{out} \cup \text{LEASAM}(s_l, Sim, \delta)$;
        $l := l + 1$;
  enddo
  return(f_{out});
end;

Note, although strata correspond to clusters, this type of stratified sampling is different from cluster sampling in statistics. Whereas ALESAM selects samples within each stratum and combines these samples to the final focusing output,

---

[9]Similarity-driven sampling covers the third disadvantage and is described separately.

cluster sampling in statistics selects a few entire cluster first and then combines these clusters to the final focusing output, or cluster sampling draws samples within each cluster again and then unifies these samples to the final focusing output.

Advantages and Disadvantages

In terms of the unifying framework of focusing solutions, ALESAM introduces two additional preparation steps before sampling. Whereas sorting is completely new to the framework, stratification is essentially a specific type of clustering which realizes a concrete *stratify* procedure in stratified sampling. Applications of leader sampling and its selection of prototypes correspond to an additional clustering step within each stratum accompanied by prototype selection. ALE-SAM does not use any explicit evaluation criterion.

In comparison to existing approaches towards focusing solutions, ALESAM is similar to PL (see algorithm 4.1.7, page 108). Stratification in ALESAM corresponds to *partition* in PL. Stratification uses the same separation strategy as PL for qualitative attributes but extends partition of tuples in case of quantitative attributes. Stratification also iteratively takes into account single attributes but utilizes pre-computed attribute relevance for the selection of the next stratification attribute rather than the more complex local quality criterion in PL. Since STRATIFY starts stratification with the class attribute as the most important attribute, it also corresponds to PL's strategy to build partitions for each class label separately. The advantage of stratification in relation to partition in PL is the more efficient approach to separate subsets of tuples into smaller strata. The usage of leader sampling in ALESAM is an alternative approach to select prototypes rather than construction of prototypes with procedure *prototype* in PL.

Now, we describe leader sampling in more detail and introduce sorting and stratification as preparation steps in ALESAM separately.

4.2.2.2 Leader Sampling

Leader sampling uses an adapted procedure of leader clustering (see algorithm 4.1.4, page 100). Instead of really building and retaining clusters, leader sampling merely stores the set of leaders and consequently saves storage requirements.

Algorithm 4.2.2 describes an implementation of leader sampling. The first tuple in the focusing input is the initial prototype in the focusing output. Afterwards, LEASAM iteratively checks for each tuple similarities to all existing prototypes in the current focusing output. If at least one prototype is more similar to the current tuple than similarity threshold δ, leader sampling does not include the new tuple into the focusing output. If no prototype in the focusing output exceeds the given threshold in comparison to the current tuple, LEASAM selects this tuple as the next prototype. At the end, LEASAM returns the final set of prototypes as the focusing output.

Algorithm 4.2.2   LEASAM(f_{in}, Sim, δ)

---

begin
  $f_{out} := \{t_1\}$;
  $i := 1$;
  while $i \leq |f_{in}|$ **do**
        $i := i + 1$;
        $i := 1$;
        while $i \leq |f_{out}|$ **do**
            if $Sim(t_{i'}, t_i) \geq \delta$
               then break;
                  else $i := i + 1$;
               endif
        enddo
        if $i > |f_{out}|$
            then $f_{out} := f_{out} \cup \{t_i\}$;
        endif
  enddo
  return(f_{out});
end;

Prototype Selection in Leader Sampling

In terms of the framework for the definition of focusing tasks, leader sampling refers to constrained subsets as the focusing output. Definition 4.2.1 states the prototype selection criterion in LEASAM. This criterion evaluates to true, if for all tuples in the focusing input, at least one tuple in the focusing output exists which exceeds similarity threshold δ in comparison to this tuple, and for all pairs of distinct tuples in the focusing output, similarities between these tuples are smaller than δ. Note, although prototype selection in LEASAM takes into account the *current* focusing output, the same condition holds for all prototypes at the end of execution. Since similarity measure Sim is symmetric, the final focusing output does not contain any prototype which is more similar to different prototypes than the specified similarity threshold.

Definition 4.2.1 (Prototype Selection in LEASAM)

Assume a (database) table (A, T), a focusing input $f_{in} \subseteq T$, and a focusing output $f_{out} \subseteq f_{in}$. Predicate $P(f_{in}, f_{out})$ in LEASAM is

$$P(f_{in}, f_{out}) \quad :\Longleftrightarrow$$
$$\forall t_i \in f_{in} \; \exists t_{i'} \in f_{out} : \quad Sim(t_i, t_{i'}) \geq \delta$$
$$\wedge \qquad \forall t_i, t_{i'} \in f_{out} : \quad t_i \neq t_{i'} \implies Sim(t_i, t_{i'}) < \delta.$$

Prototype selection in LEASAM extends the prototype selection criterion in IBL2 (see definition 4.1.8, page 106). Whereas IBL2 considers information of all attributes to retrieve the most similar tuple in the current focusing output but only uses information on the class attribute to decide whether to retain tuples or not, leader sampling considers more information since prototype selection takes into account overall similarities to decide whether to add new tuples to the focusing output or not. Moreover, similarity threshold δ makes LEASAM more flexible in adjusting the selection criterion in different focusing contexts.

4.2.2.3 Sorting

Recall, the first drawback of LEASAM is its sensitivity to the order of tuples within the focusing input. For each particular order of tuples, leader sampling is likely to generate different focusing outputs. If we add a preparation step to LEASAM which arranges tuples in a fixed well-defined manner, succeeding leader sampling is independent of the original order in the focusing input. Then, leader sampling is still sensitive to the order of tuples, but for different sequences of tuples in the same original focusing input, LEASAM always generates the same focusing output. This makes leader sampling deterministic, and the final focusing output is less randomly created. Note, since this drawback is not only a specific disadvantage of leader sampling but many techniques suffer from their order-sensitivity, a mechanism to overcome this negative property is also beneficial for other approaches.

Order Relation $\succ$

For these reasons, we define a specific sorting criterion in order to re-arrange the focusing input into a fixed well-defined order before leader sampling starts. The key idea of this sorting criterion is to re-arrange all tuples in the focusing input such that more similar tuples are more likely to be close to each other in the sorted table. Therefore, we define the following order relation (see definition 4.2.2).

Definition 4.2.2 (Order Relation $\succ$)

Assume a (database) table (A, T), *attribute relevance weights* $w_j \geq 0$, $1 \leq j \leq N$, *and w.l.o.g.* $w_N \geq w_{N-1} \geq \ldots \geq w_j \geq \ldots \geq w_1$. *We define the order relation* $\succ \subseteq T \times T$ *as*

$$(i) \qquad \succ (t_i, t_{i'}) \quad :\Longleftrightarrow$$
$$\succ (t_{i[1;\ N]}, t_{i'[1;\ N]}),$$

$$(ii) \quad \succ (t_{i[1;\ j]}, t_{i'[1;\ j]}) \quad :\Longleftrightarrow$$
$$t_{ij} > t_{i'j} \lor \left(t_{ij} \equiv t_{i'j} \land \succ (t_{i[1;\ j-1]}, t_{i'[1;\ j-1]}) \right),$$

$$(iii) \quad \not\succ (t_{i[1;\ 0]}, t_{i'[1;\ 0]})$$

If $\succ (t_i, t_{i'})$, we denote the first j, $N \geq j \geq 1$, which meets the first condition of (ii) by $j_{ii'}^*$, i.e., $j_{ii'}^* := \max_{j=1}^{N}\{j \mid t_{ij} > t_{i'j}\}$.

Without loss of generality, we assume that the sequence of attributes in the attribute set reflects attribute relevance in increasing order. If the value for the most important attribute in a tuple is larger than the corresponding value in another tuple, the index of the first tuple in the sorted table is larger than the position of the other tuple. If both tuples have the same value for an attribute, sorting recursively considers the next important attribute in the same way. If two tuples are completely identical, the relation between their locations in the sorted table is the same as in the original table. For the order relation $>$ of attribute values, we use the numeric order for quantitative attributes and the lexicographic order for qualitative attributes. The following proposition shows that the resulting relation is a well-defined order.

Proposition 4.2.1 (Order Relation $\succ$)

Assume a (database) table (A, T). Relation $\succ \subseteq T \times T$ is a well-defined order.

Proof:

$\succ \subseteq T \times T$ meets the following conditions of a well-defined order:

(i)   Irreflexivity:
$$\forall t_i \in T : \quad \not\succ (t_i, t_i),$$

(ii)   Transitivity:
$$\forall t_i, t_{i'}, t_{i''} \in T : \quad \succ (t_i, t_{i'}) \wedge \succ (t_{i'}, t_{i''}) \implies \succ (t_i, t_{i''}),$$

(iii)   Asymmetry:
$$\forall t_i, t_{i'} \in T : \quad \succ (t_i, t_{i'}) \implies \not\succ (t_{i'}, t_i),$$

(iv)   Konnexity:
$$\forall t_i, t_{i'} \in T : \quad \succ (t_i, t_{i'}) \vee \succ (t_{i'}, t_i) \vee t_i \equiv t_{i'}.$$

Thus, $\succ \subseteq T \times T$ is a well-defined order. $\square$

Similarity in Sorted Tables

If we apply similarity measure Sim (see definition 2.5.9, page 40), order relation $\succ$ groups locally more similar tuples together. Since sorting starts with the most important attribute and considers attributes with decreasing relevance in case of ties, it is also likely that overall more similar tuples are positioned close to each other in the sorted table.

Proposition 4.2.2 (Local Similarity in Sorted Tables)

Assume a (database) table (A, T).

$\forall t_i, t_{i'}, t_{i''} \in T :$

$\succ (t_i, t_{i'}) \wedge \succ (t_{i'}, t_{i''}) \quad \Longrightarrow$

(i) $j_{ii'}^* \leq j_{ii''}^*$, and

(ii) $sim_j(t_{ij}, t_{i'j}) \geq sim_j(t_{ij}, t_{i''j}) \; \forall \, N \geq j \geq j_{ii'}^* + 1$

Proof:

(i) *We assume $j_{ii'}^* > j_{ii''}^*$. Then, by definition 4.2.2 $t_{iN} = t_{i'N}$, $t_{iN-1} = t_{i'N-1}$, ..., $t_{ij*_{ii'}+1} = t_{i'j*_{ii'}+1}$, and $t_{ij*_{ii'}} > t_{i'j*_{ii'}}$, as well as $t_{iN} = t_{i''N}$, $t_{iN-1} = t_{i''N-1}$, ..., $t_{ij*_{ii'}} = t_{i''j*_{ii'}}$, ..., $t_{ij*_{ii''}+1} = t_{i''j*_{ii''}+1}$, and $t_{ij*_{ii''}} > t_{i''j*_{ii''}}$.*

This means $t_{iN} = t_{i'N} = t_{i''N}$, $t_{iN-1} = t_{i'N-1} = t_{i''N-1}$, ..., $t_{ij*_{ii'}+1} = t_{i'j*_{ii'}+1} = t_{i''j*_{ii'}+1}$, and $t_{ij*_{ii'}} = t_{i''j*_{ii'}} > t_{i'j*_{ii'}}$, i.e., $\succ (t_{i''}, t_{i'})$ in contradiction to $\succ (t_{i'}, t_{i''})$ and the asymmetry of $\succ$.

(ii) *follows from definition 4.2.2.* □

Proposition 4.2.2 provides two auxiliary characteristics of sorted tables in order to analyze similarities in sorted tables. First, if the position of a tuple is smaller than the position of another tuple which is again placed before a third tuple in sorted tables, then the number of identical attribute values for the most relevant attributes between the first two tuples is not smaller than the respective number between the first and third tuple. Second, local similarities between values of the first two tuples are higher than or equal to the same local similarities between values of the first and third tuple for all attributes which are more relevant than the first attribute that determines the relation between the first two tuples.

Now, we use both conditions in proposition 4.2.2 to show properties of similarities in sorted tables. Proposition 4.2.3 compares overall similarities among tuples in sorted tables. Note, although order relation $\succ$ is likely to place more similar tuples close to each other, this assumption is not true in general. However, given an additional precondition, proposition 4.2.3 shows that tuples in sorted tables are more similar, if the distance between their positions decreases.

Proposition 4.2.3 (Similarity in Sorted Tables)

Assume a (database) table (A, T).

$\forall t_i, t_{i'}, t_{i''} \in T:$

$$\succ (t_i, t_{i'}) \ \wedge \ \succ (t_{i'}, t_{i''}) \ \wedge \ \sum_{j=1}^{j^*_{ii'}} w_j \cdot sim_j(t_{ij}, t_{i'j}) \geq \sum_{j=1}^{j^*_{ii'}} w_j \cdot sim_j(t_{ij}, t_{i''j})$$

$$\implies \quad Sim(t_i, t_{i'}) \ > \ Sim(t_i, t_{ii''})$$

Proof:

See appendix C, proposition C.0.2, page 278.   □

Since sorting is likely to arrange similar tuples close to each other, order relation $\succ$ is particularly useful for leader sampling, because leader sampling considers more similar tuples in sequences such that prototype selection returns more appropriate representations of subsets of sufficiently similar tuples.

If we take into account selection of split attributes in top down induction of decision trees, we expect that sorting in combination with leader sampling is especially beneficial for quantitative attributes. Sorting ensures that leader sampling considers minimum values for important quantitative attributes first and is then likely to select additional tuples which represent maximum values for the same attribute later. This is advantageous for the determination of threshold values for quantitative attributes and the examination of attributes with respect to information gain ratio in top down induction of decision trees. For similar reasons, sorting is also a beneficial enhancement to systematic sampling and decreases effects of randomness in statistical sampling techniques.

4.2.2.4 Stratification

Stratification is the second improvement in ALESAM in order to overcome the second major drawback in leader sampling. Stratification provides an additional preparation step that follows up the sorting approach. Stratification constructs subsets of locally similar tuples according to attribute values, and each subset corresponds to a stratum. Similar to sorting, this construction process considers attributes in decreasing order of relevance. Consequently, stratification is also likely to group overall similar tuples into strata.

The goal of stratification is two-fold. First, applications of more intelligent, and hence more expensive, sampling is more efficient, if we process each stratum separately and combine resulting subsamples to the final focusing output. In particular, this approach makes advanced leader sampling more feasible than pure LEASAM, if we assume that costs for stratification are low in comparison to leader sampling on the entire focusing input. Second, stratification is also often recommended in statistics. For appropriate stratification procedures, variances within strata decrease, and probability distributions in the unified sample are more likely to approximately represent probability distributions in the original focusing input. In particular, for estimations of proportions, stratified sampling yields lower variances.

The stratification procedure proposed here separates tuples into strata according to their attribute values for single attributes. If we only consider one attribute, this stratification procedure is local. Overall stratification recursively applies local stratification to each stratum until a termination criterion holds. Stratification considers attributes, one by one, in order of decreasing relevance. The more relevant an attribute is, the earlier stratification takes into account this attribute in local stratification. In this aspect, stratification follows up the sorting approach.

Local Stratification

Local stratification distinguishes between qualitative and quantitative attributes. For qualitative attributes, local stratification separates tuples into exhaustive and non-overlapping strata such that each tuple has the same value for the considered attribute within each stratum. Definition 4.2.3 characterizes local strata for qualitative attributes.

Definition 4.2.3 (Local Strata for Qualitative Attributes)

Assume a (database) table (A, T)*, a focusing input* $f_{in} \subseteq T$*, and a qualitative attribute* $a_j \in A$ *with domain* $dom(a_j) = \{a_{j1}, \ldots, a_{jk}, \ldots, a_{jN_j}\}$*. We define a local stratum* s_{jk} *in* f_{in} *for* a_j *as*

$$s_{jk} := \{t_i \in f_{in} \mid t_{ij} \equiv a_{jk}\}.$$

If we follow the same approach for quantitative attributes, i.e., build strata for each occurring value in an attribute domain, stratification tends to result in many small strata. Consequently, we automatically get larger focusing outputs, if we apply sampling techniques to each stratum separately and combine the resulting samples. Since this approach is less appropriate for focusing, we first group similar values for quantitative attributes and then consider different values in each group as the same value in the sense of qualitative attributes. For each group of values, local stratification generates strata similar to the procedure for qualitative attributes. Within each stratum, each tuple represents an attribute value that belongs to the same group of values.

Definition 4.2.4 (Local Strata for Quantitative Attributes)

Assume a (database) table (A, T)*, a focusing input* $f_{in} \subseteq T$*, and a quantitative attribute* $a_j \in A$ *with domain* $dom(a_j) \subseteq I_{j1} \cup \ldots \cup I_{jl} \cup \ldots \cup I_{jL_j}$ *and* $I_{jl} \cap I_{jl'} = \emptyset \ \forall l, l', \ 1 \leq l, l' \leq L_j, \ l \neq l'$*. We define a local stratum* s_{jl} *in* f_{in} *for* a_j *as*

$$s_{jl} := \{t_i \in f_{in} \mid t_{ij} \in I_{jl}\}.$$

Since quantitative attribute domains are subsets of real values, groups of similar values typically correspond to intervals specified by lower and upper boundaries. *Discretization* approaches to build groups of similar values as intervals are discussed separately in subsection 4.2.2.5. Definition 4.2.4 assumes the existence of intervals and states the concept of local strata for quantitative attributes more precisely.

Stratify

The overall stratification procedure starts local stratification with the most important attribute and then recursively applies local stratification to each stratum according to the next important attribute in decreasing order of relevance. Stratification iterates local stratification until a termination criterion holds. Here, we use a threshold for the maximum number of tuples within a stratum as the termination criterion. Stratification does not continue local stratification, if the size of the current stratum does not exceed the pre-defined threshold.

Procedure 4.2.1 shows an implementation of overall stratification in advanced leader sampling. First, STRATIFY initializes the set of strata. If stratification has processed all attributes or observed the first irrelevant attribute (which means that only irrelevant attributes follow), it returns the current stratification input since no more (relevant) attributes are available for local stratification. Otherwise, stratification initializes the number of strata. In case of qualitative attributes, the number of strata corresponds to the number of values in the attribute domain. In case of quantitative attributes, stratification builds groups of similar values by applying discretization to the attribute domain, and the number of resulting intervals defines the number of strata. Note, we describe specific examples for procedure *discretize* and its parameters in the next section separately.

After initialization of each stratum to an empty set, stratification computes the stratum index for each tuple in the current stratification input. Again, we need to distinguish between qualitative and quantitative attributes. If the current stratification attribute is qualitative, the stratum index of a tuple is the index of its respective attribute value.[10] In case of quantitative stratification attributes, stratification seeks for the index of the interval that contains the tuple's attribute value. The index of this interval is then the index of this tuple's stratum.

After stratification processed all tuples in the current stratification input, it checks the resulting set of strata. If the number of tuples in a stratum exceeds the pre-specified threshold max, this stratum is again stratified according to the next important stratification attribute. Otherwise, stratification adds the stratum to the set of final strata. At the end, stratification returns the final set of strata.

---

[10]We assume that we refer to values in qualitative attribute domains by their index.

Procedure 4.2.1   STRATIFY($f_{in}, j, max, discretize$)

```
begin
  S_j := ∅;
  if j = 0 ∨ w_j = 0
    then return(f_in);
  endif
  if a_j qualitative
    then L := N_j;
  endif
  if a_j quantitative
    then I_j := discretize(dom(a_j), ...);
         L := | I_j |;
  endif
  l := 1;
  while l ≤ L do
      s_jl := ∅;
      l := l + 1;
  enddo
  i := 1;
  while i ≤ | f_in | do
      if a_j qualitative
        then s_{jt_{ij}} := s_{jt_{ij}} ∪ {t_i};
      endif
      if a_j quantitative
        then l := 1;
            while l ≤ L do
                if t_{ij} ∈ I_{jl}
                  then s_jl := s_jl ∪ {t_i};
                       break;
                  else l := l + 1;
                endif
            enddo
      endif
  enddo
  l := 1;
  while l ≤ L do
      if | s_jl | > max
        then S_j := S_j ∪ STRATIFY(s_jl, j - 1, max, discretize);
        else S_j := S_j ∪ s_jl;
      endif
      l := l + 1;
  enddo
  return(S_j);
end;
```

In comparison to stratification in ALESAM, procedure 4.2.1 contains two simplifications. First, this procedure ignores unknown values. In ALESAM, stratifi-

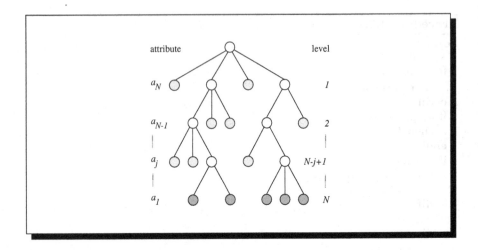

Figure 4.3: Strata Tree

cation treats unknown values as an extra value in the current attribute domain. Hence, local stratification creates an additional stratum and puts all tuples with unknown values for the current stratification attribute into this stratum.

Second, the test loop at the end of procedure 4.2.1 in ALESAM also checks, if an initially empty stratum remains empty after assignments of tuples to strata. Since stratification recursively applies local stratification with different stratification attributes, it is possible that all tuples with certain values already belong to different strata at higher levels of relevance. Thus, empty strata possibly occur. Since it does not make sense to proceed with empty strata, stratification removes empty strata.

The implementation of stratification in ALESAM also contains additional termination criteria that support alternatives to the pre-specified maximum stratum size. Stratification is also able to stop local stratification after consideration of a pre-specified number of attributes. For example, if we set the number of attributes to 3, stratification considers only the three top most relevant attributes for local stratification. Stratification in ALESAM supplies two options, either to use only the most important attribute or to take into account all relevant attributes, regardless of the local stratum size.

Strata Trees and Clustering

Figure 4.3 illustrates stratification and resulting *strata trees*. At each tree level, local stratification considers a single attribute a_j. Each node corresponds to a stratum and is separated into smaller strata according to the local stratification process, if this stratum still contains too many tuples. Stratification recursively proceeds with the next important attribute until the stratum size is smaller than the pre-defined threshold (light gray nodes in figure 4.3), or the last available stratification attribute has been processed (dark gray nodes in figure 4.3). At the end, stratification returns all leafs of the resulting strata tree. Note, this illustration assumes positive attribute relevance weights for all attributes.

Definition 4.2.5 (Stratification)

Assume a (database) table (A, T), a focusing input $f_{in} \subseteq T$, and L sets of tuples $S = \{s_1, s_2, \dots, s_l, \dots, s_L\}$. S is stratification of f_{in}, if

(i) $\displaystyle\bigcup_{l=1}^{L} s_l = f_{in}$,

(ii) $\forall s_l, s_{l'} \in S, \ l \neq l' : \ s_l \cap s_{l'} = \emptyset$,

(iii) $\forall s_l \in S \ \exists a_j \in A :$

s_l *is local stratum in* f_{in} *for* a_j, *or* $\exists s_{l'} \in S : \ s_l$ *is local stratum in* $s_{l'}$ *for* a_j *and* $s_{l'}$ *is local stratum in* f_{in} *for* a_{j+1}.

Definition 4.2.5 summarizes characteristics of stratifications generated with procedure 4.2.1. The resulting set of strata is essentially an exhaustive, non-overlapping, and non-hierarchical clustering since unification of all strata equals the focusing input, intersections of pairs of distinct strata are empty, and each stratum is either a local stratum in the focusing input or a local stratum in its preceding strata within the strata tree.[11]

Proposition 4.2.4 (STRATIFY and Clustering)

Assume a (database) table (A, T), a focusing input $f_{in} \subseteq T$, and $max \geq 1$. Then, STRATIFY$(f_{in}, N, max, discretize)$ is an exhaustive, non-overlapping, non-hierarchical clustering of f_{in}.

Proof:

Assume $S := $ STRATIFY$(f_{in}, N, max, discretize) = \{s_1, s_2, \dots, s_l, \dots, s_L\}$. Then, S meets the following conditions:

(i) $\forall s_l \in S : \ s_l \neq \emptyset$ *and* $s_l \subseteq f_{in}$,

(ii) $\forall t_i \in f_{in} \ \exists s_l \in S : \ t_i \in s_l$, *and*

(iii) $\forall s_l, s_{l'} \in S, \ l \neq l' : \ s_l \cap s_{l'} = \emptyset$.

This means by definitions 4.1.3, 4.1.5, 4.1.6, and 4.1.7 that STRATIFY$(f_{in}, N, max, discretize)$ is an exhaustive, non-overlapping, non-hierarchical clustering of f_{in}. □

---

[11]Definitions 4.2.3 and 4.2.4 only describe local strata at the top level. The definition of local strata at lower levels is identical except that tuples are now elements of the preceding stratum.

Proposition 4.2.4 shows the relation between stratification and clustering in more detail. Note, although original strata trees correspond to hierarchical clusterings, we continue the focusing process only with leaves in these hierarchies.

Strata Trees and Sorting

In some sense, stratification supersedes sorting. If we first sort the focusing input according to order relation $\succ$ (see definition 4.2.2, page 117), we are able to implement stratification by marking boundaries of strata in the sorted table. We generate the first strata level according to the most important attribute by marking the end of each subset in the sorted focusing input when the value (or the interval of values) changes. We recursively replicate this marking procedure according to the next important attribute within each pair of boundaries from the previous step. If we stop the local marking process as soon as the set of tuples between two boundaries contains less than a pre-specified number of tuples or no more (relevant) attributes are available, the set of subsets of tuples between each pair of boundaries constitutes the same set of strata as leaves in the resulting strata tree in procedure 4.2.1. Consequently, propositions 4.2.2 and 4.2.3 on similarity in sorted tables also apply to similarity within strata in analogous ways.

Strata Trees and kd-Trees

The similarity between stratification and applications of *kd-trees* in case-based reasoning (Bentley, 1975; Wess *et al.*, 1994) indicates the specific usefulness of stratification as a preparation step before leader sampling in ALESAM. Strata trees are comparable to kd-trees which separate tuples into hierarchies of subsets in similar ways.

Local separation of tuples according to single attributes is identical for both, strata trees and kd-trees, if the current attribute is qualitative. In contrast, construction of local strata in case of quantitative attributes is slightly different, if we compare strata trees and kd-trees. Whereas procedure 4.2.1 utilizes discretization approaches that often result in more than two intervals, kd-trees only allow two intervals according to a single partition value. Likewise, procedure 4.2.1 never considers single attributes again at lower levels, whereas kd-trees possibly use the same (quantitative) attribute more than once.

Strata trees and kd-trees also differ in their strategy to select the next attribute for local separation. Procedure STRATIFY uses a fixed pre-defined order of attributes according to decreasing relevance and computes attribute relevance on the entire focusing input before stratification. In contrast, kd-trees compute the best split attribute locally according to maximum variations of values in attribute domains. Finally, the maximum stratum size in procedure 4.2.1 corresponds to the bucket size in kd-trees.

Although the purpose of kd-trees in case-based reasoning is different, the analogy between strata trees and kd-trees motivates usability of stratification in ALESAM. In case-based reasoning, kd-trees enable faster retrieval of most similar cases for given query cases without reducing the number of available tuples for classification. Two tests, called *ball within bounds* (BWB) and *ball overlaps bounds* (BOB), decide whether to search additional nodes in the tree, in order to find the most similar case, and recommend at which node the search proceeds. Hence, case-based reasoning systems with kd-trees for retrieval often do not need to consider all tuples in order to find most similar cases for given query cases. Extensions of BWB and BOB tests which utilize *dynamic* bounds further speed up retrieval of most similar cases in case-based reasoning, since they reduce the amount of unnecessary bounds tests and their partly too pessimistic results (Wess, 1995).

Applications and advantages of kd-trees in case-based reasoning justify stratification in advanced leader sampling in the following way. The correspondence between strata trees and kd-trees implies that results in case-based reasoning also hold for stratification in ALESAM. Consequently, stratification efficiently supports retrieval of most similar prototypes for given tuples in leader sampling.

If we integrate BWB and BOB tests in ALESAM, we are able to ensure that we definitely find the most similar prototype in the entire focusing input for each tuple. Since failing this only means that tuples are possibly selected as prototypes in the final focusing output although another tuple which is more similar and exceeds similarity threshold δ in comparison to these prototypes eventually exists in a different stratum, we only retain more tuples in the focusing output than really necessary according to the prototype selection criterion in LEASAM. On the other hand, if we implement BWB and BOB tests in advanced leader sampling, computational costs in ALESAM increase. Thus, we decide not to integrate BWB and BOB tests, in order to keep focusing solutions more efficient.

Further Usage of Stratification

Although stratification is primarily used in advanced leader sampling, we also apply the same stratification procedure in combination with different sampling techniques. For example, if we utilize simple random sampling instead of leader sampling within each stratum, the combination of stratification and simple random sampling results in a particular stratified simple random sampling approach (see section 4.1.2.3, page 90). In this case, we compute sample sizes within each stratum proportional to relative sizes of strata in the focusing input. Since selection costs for each tuple within each stratum are identical, this allocation strategy is close to optimum allocation in statistics but computationally less expensive (see Cochran (1977), page 97 and page 109).

For the selected classification algorithms, we expect that stratification is also advantageous. Since top down induction of decision trees uses information gain

criteria for selections of the best split attribute, and information gain relies on
proportions, stratification is suitable, because stratified sampling yields close
estimations of proportions, if local construction of strata places tuples with
identical values into the same stratum (see Cochran (1977), page 107). This
corresponds exactly to local stratification in procedure 4.2.1. Likewise, since
stratification groups similar tuples and we draw separate samples within each
stratum, the final focusing output contains dissimilar tuples which represent the
entire focusing input and consequently cover the most important information for
nearest neighbor classifiers.

4.2.2.5 Discretization

Now, we return to the question of identifying groups of similar values within
domains of quantitative attributes which we use in stratification. For this purpose, we apply *discretization* approaches which have been proposed in the past
in different variations. In general, discretization builds non-overlapping intervals
that cover the entire attribute domain of quantitative attributes. An interval
is a range of quantitative values defined by its lower and upper boundary values. We identify intervals in discretization of attribute domains such that end
values of intervals refer to start values of succeeding intervals, and the lower
boundary of the first interval is the minimum value among occurring attribute
values, whereas the upper boundary of the last interval is the maximum value
among occurring attribute values, respectively. Definition 4.2.6 precisely states
requirements for discretizations of quantitative attribute domains.

Definition 4.2.6 (Discretization)

*Assume a (database) table (A, T), a quantitative attribute $a_j \in A$, and a set of
intervals $I = \{I_1, I_2, \ldots, I_l, \ldots, I_{L_j}\}$ with $I_l = [b_l, e_l[$, $b_l \leq e_l$, $1 \leq l \leq L_j - 1$,
and $I_{L_j} = [b_{L_j}, e_{L_j}]$. I is discretization of $dom(a_j)$, if*

$$
\begin{array}{rll}
(i) & dom(a_j) & \subseteq \ \displaystyle\bigcup_{l=1}^{L_j} I_l, \\[2ex]
(ii) & b_1 & = \ \displaystyle\min_{k=1}^{N_j}(dom(a_j)), \\[2ex]
(iii) & e_{L_j} & = \ \displaystyle\max_{k=1}^{N_j}(dom(a_j)), \text{ and} \\[2ex]
(iv) & e_l & = \ b_{l+1}, \ 1 \leq l \leq L_j - 1.
\end{array}
$$

Discretization Strategies

Dougherty *et al.* (1995) propose three different aspects for characterizations
of discretization strategies:

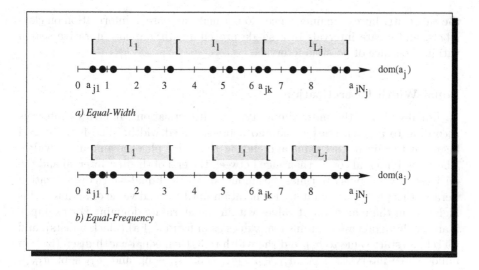

Figure 4.4: Equal-Width and Equal-Frequency Discretization

- *Global and Local Discretization*

  Whereas global discretization considers the entire focusing input at once and creates intervals by taking into account all occurring values at the same time, local approaches decompose the focusing input into local regions and construct intervals for each region separately.

- *Unsupervised and Supervised Discretization*

  Unsupervised discretization considers only information available in the set of attribute values for single attributes and does not make any use of information on class labels. In contrast, supervised approaches also consider class information and try to build intervals that gain maximum benefit with respect to classification goals.

- *Static and Dynamic Discretization*

  Similar to the distinction in evaluation criteria for focusing success, discretization approaches differ whether they utilize static procedures and fixed numbers of intervals, or whether they conduct dynamic searches over possible values for the number of intervals and try to find the most appropriate number of intervals during discretization.

Although local and supervised discretization approaches are often superior in comparison to global and unsupervised discretization techniques, and it also seems reasonable to search for the most appropriate number of intervals, we decide to keep discretization in advanced leader sampling as simple as possible. Figure 4.4 illustrates two global, unsupervised, and static discretization techniques. *Equal-width* and *equal-frequency* discretization both take into account the entire attribute domain at once, do not make any use of information on class labels, and create intervals in a single procedure without any iterative search within the space of possible intervals.

Equal-Width Discretization

Equal-width is the most simple type of discretization. It creates intervals according to a given start value and a pre-specified width. The first interval begins at the given start value and ends at this value plus the amount of width. The second interval covers the range between the end of the first interval and the initial start value plus two times the amount of width. Equal-width discretization iterates this procedure of interval generation until the end value of an interval is higher than the maximum of values within the attribute domain. For example, if an attribute domain contains real values as in figure 4.4 a) (black bullets), and we set the start value to a_{j1} and the width to 3, then equal-width discretization builds three intervals: $[a_{j1};\ a_{j1} + 3[,\ [a_{j1} + 3;\ a_{j1} + 6[$, and $[a_{j1} + 6;\ a_{jN_j}]$. Note, in order to meet the definition of discretization (see definition 4.2.6, page 128), we set the end value of the last interval to the maximum value in domain $dom(a_j)$.

Procedure 4.2.2 $e\_w(j,L)$

---

begin
   $I := \emptyset$;
   $b_1 := \min\limits_{k=1}^{N_j}(dom(a_j))$;
   $e_L := \max\limits_{k=1}^{N_j}(dom(a_j))$;
   $width := \dfrac{e_L - b_1}{L}$;
   $l := 1$;
   while $l \leq L - 1$ **do**
        $e_l := b_l + width$;
        $b_{l+1} := e_l$;
        $I := I \cup [b_l, e_l[$;
        $l := l + 1$;
   enddo
   $I := I \cup [b_L, e_L]$;
   return(I);
end;

Procedure 4.2.2 illustrates an implementation of equal-width discretization. First, this procedure initializes the set of intervals to an empty set, specifies the first start value and the last end value of intervals as the minimum and maximum value in attribute domain $dom(a_j)$, respectively, and computes the size of *width* according to minimum and maximum value in $dom(a_j)$ and a pre-defined number of intervals L. Then, procedure 4.2.2 iteratively builds L intervals of equal width in attribute domain $dom(a_j)$ such that the end of an interval always refers to the start of the succeeding interval. Note, this implementation differs from the description of equal-width discretization above, since it computes the value of width for a pre-specified number of intervals instead of using a pre-defined width directly. This is advantageous, if we know the number of intervals that we want to achieve, but have no information about occurring values in attribute domain $dom(a_j)$.

Equal-Frequency Discretization

The main disadvantage of equal-width discretization is the possibility to generate unbalanced intervals. Whereas some intervals possibly cover many values occurring in the database, other intervals eventually only contain a small number of values. Equal-frequency discretization overcomes this potential drawback and tries to balance the number of tuples with values within each interval.

In equal-frequency discretization, we either specify the number of tuples within each interval or the number of intervals. If we specify the number of tuples, equal-frequency discretization iteratively adds values to an interval until the interval covers the pre-defined number of tuples. Then, it starts a new interval and proceeds in the same way. If we specify the number of intervals instead, equal-frequency discretization separates tuples and their values in attribute domain $dom(a_j)$ into a number of equally sized subsets, which corresponds to the number of intervals, and defines an interval for the range of values within each of these subsets. Figure 4.4 b) illustrates equal-frequency discretization for a pre-specified number of tuples 5. Note, the last interval only covers 4 tuples since attribute domain $dom(a_j)$ does not contain more values.

If we take into account the purpose of discretization in advanced leader sampling, specification of the number of tuples within each interval is more appropriate since stratification uses a maximum number of tuples allowed within each stratum as its termination criterion.

Procedure 4.2.3 shows an implementation of equal-frequency discretization for pre-specified numbers of tuples within each interval. This implementation assumes that the set of occurring values for attribute a_j in the focusing input is sorted in advance. Equal-frequency discretization starts with initialization of the set of intervals to an empty set, sets the start of the first interval to the jth value of the first tuple (which corresponds to the minimum value in domain $dom(a_j)$ since we assume sorted values), and initializes the index for tuples as well as the counter for the number of intervals to 1.

While discretization has not considered all tuples in the focusing input, it increments index i by the pre-specified number of tuples L. If the new position still refers to an existing tuple in the focusing input, equal-frequency discretization sets the end value of the next interval to the jth attribute value of the ith tuple. Otherwise, the end of the next interval equals the jth value of the last tuple in the focusing input. This means that the last interval gets the rest of occurring values which possibly corresponds to less than L tuples.

Procedure 4.2.3 $e\_f(f_{in}, j, L)$

---
begin
  $I := \emptyset;$
  $b_1 := t_{1j};$
  $i := 1;$
  $l := 1;$
  while $i \ < \mid f_{in} \mid$ **do**
      $i := i + L;$
      if $i \ \leq \mid f_{in} \mid$
        then $e_l := t_{ij};$
          else $e_l := t_{\mid f_{in} \mid j};$
      endif
      while $b_l = e_l \ \wedge \ i+1 \ \leq \mid f_{in} \mid$ **do**
          $i := i + 1;$
          $e_l := t_{ij};$
      enddo
      if $i \ < \mid f_{in} \mid$
        then $I := I \cup [b_l, e_l[;$
          else $I := I \cup [b_l, e_l];$
      endif
      $b_{l+1} := e_l;$
      $l := l + 1;$
  enddo
  return$(I);$
end;

In order to avoid having identical start and end values of an interval, equal-frequency discretization seeks for new end values as long as start and end values of an interval coincide and more tuples remain in the focusing input. Then, procedure 4.2.3 adds the resulting interval to the set of current intervals and initializes the start value of the next interval to the end value of the previous interval. In this aspect, this implementation differs from the illustration in figure 4.4 b). Whereas the illustration contains empty regions between intervals, this implementation of equal-frequency discretization ensures that the final set of intervals meets the definition of discretization.

Properties of Equal-Width and Equal-Frequency Discretization

Proposition 4.2.5 summarizes that both implementations of equal-width and equal-frequency discretization generate discretizations according to definition 4.2.6.

Proposition 4.2.5 (Discretization)

Assume a (database) table (A, T), a focusing input $f_{in} \subseteq T$, a quantitative attribute $a_j \in A$, and $L \geq 1$. Then, $e\_w(j, L)$ and $e\_f(f_{in}, j, L)$ are discretizations of $dom(a_j)$.

Proof:

Assume $I := e\_w(j, L) = \{I_1, I_2, \ldots, I_l, \ldots, I_L\}$ with $I_l = [b_l, e_l[$, $b_l \leq e_l$, $1 \leq l \leq L - 1$, and $I_L = [b_L, e_L]$, and $I' := e\_f(f_{in}, j, L) = \{I'_1, I'_2, \ldots, I'_l, \ldots, I'_{L'}\}$ with $I'_l = [b'_l, e'_l[$, $b'_l \leq e'_l$, $1 \leq l \leq L' - 1$, and $I'_{L'} = [b'_{L'}, e'_{L'}]$. Then, I and I' meet the following conditions:

$$(i) \quad dom(a_j) \subseteq \bigcup_{l=1}^{L} I_l,$$

$$dom(a_j) \subseteq \bigcup_{l=1}^{L'} I'_l,$$

$$(ii) \quad b_1 = \min_{k=1}^{N_j}(dom(a_j)),$$

$$b'_1 = \min_{k=1}^{N_j}(dom(a_j)),$$

$$(iii) \quad e_L = \max_{k=1}^{N_j}(dom(a_j)),$$

$$e'_{L'} = \max_{k=1}^{N_j}(dom(a_j)),$$

$$(iv) \quad e_l = b_{l+1}, \ 1 \leq l \leq L - 1,$$

$$e'_l = b'_{l+1}, \ 1 \leq l \leq L' - 1.$$

This means by definition 4.2.6 that $e\_w(j, L)$ and $e\_f(f_{in}, j, L)$ are discretizations of $dom(a_j)$. □

Discretization in Advanced Leader Sampling

Due to the termination criterion in stratification, we recommend applying equal-frequency discretization in advanced leader sampling rather than equal-width discretization. Equal-frequency discretization allows more control on the number of tuples with values within each interval than equal-width discretization. Note, stratification in ALESAM computes the number of tuples L for equal-frequency discretization before local stratification by computing values that balance the number of remaining attributes for stratification and the maximum number of tuples allowed within each stratum. If stratification is still at high levels in the strata tree, i.e., many attributes remain for local stratification, it sets the number of tuples L to higher values than *max* to ensure that stratification is also able to make use of subsequent attributes and does not stop too early at an inappropriate high level of generality. The higher the current level in stratification is, i.e., the more attributes are still available for local stratification, the larger is the difference between the current value for the number of tuples L in equal-frequency discretization and the maximum number of tuples allowed within each stratum.

Alternative Discretization Approaches

In addition to equal-width and equal-frequency discretization, we also implemented and tested two supervised and dynamic discretization strategies (Fayyad & Irani, 1993; Hong, 1994). Initial experiments showed that both methods are more appropriate in some situations. However, these discretization approaches highly affect running time, and stratification becomes less efficient. Since we are interested in the development of efficient focusing solutions, we do not make further usage of these techniques.

4.2.3 Similarity-Driven Sampling

Similarity-driven sampling further extends advanced leader sampling in several ways:

- First, we develop similarity-driven automatic estimations and adaptations of similarity threshold δ in leader sampling since success of LEASAM mostly relies on appropriate settings of this threshold. Thereby, we overcome the third major drawback of pure leader sampling.

- Second, we propose an additional strategy to speed up leader sampling. Similarity-driven sampling no longer processes the entire focusing input if not necessary. We regard consideration of the entire focusing input as not necessary, if sampling does not make any progress on varying systematic samples of a pre-specified size for a number of iterations. This strategy further contributes to overcome possibilities of many similarity comparisons in leader sampling, if the focusing output becomes large.

- Third, we introduce prototype weights to count the number of tuples that each prototype represents. Prototype weights consolidate information on the importance of each prototype in the focusing output.

- Finally, similarity-driven sampling retrieves the most similar prototype in the current focusing output instead of stopping, if the first prototype exceeds similarity threshold δ. This is important in combination with prototype weights, in order to reflect the level of representativeness of each prototype properly. We denote this modified version of LEASAM by a superscript $+$.[12]

---

[12]Note, if the similarity between a new tuple and its nearest prototype exceeds similarity threshold δ if and only if class labels of both coincide, LEASAM$^+$ corresponds to IBL2 (see algorithm 4.1.6, page 105).

In the following, we first outline the similarity-driven sampling approach. Then, we develop heuristics to automatically estimate and adapt similarity threshold δ. Afterwards, we describe the integration of prototype weights in similarity-driven sampling in order to eliminate less representative prototypes in the focusing output. Finally, we return to attribute relevance and propose several measures to compute attribute relevance weights automatically.

4.2.3.1 Algorithm SIMSAM

Algorithm 4.2.3 presents an implementation of similarity-driven sampling. After initialization of the focusing output to an empty set, SIMSAM initially estimates similarity threshold δ, sets the progress counter to 0 and the start position to 1, and computes the step size for systematic sampling according to the focusing input size and a pre-specified number of tuples for each intermediate working set. Then, SIMSAM starts the first sampling loop.

Algorithm 4.2.3   SIMSAM$(f_{in}, m, Sim, adapt, size, iterations)$

begin
  $f_{out} := \emptyset$;
  $\delta := initialize(f_{in}, adapt)$;
  $progress := 0$;
  $start := 1$;
  $step := \lfloor \frac{|f_{in}|}{size} \rfloor$;
  while $(start < step) \wedge (start - progress < iterations)$ **do**
      $m := |f_{out}|$;
      $f_{out} := f_{out} \cup$ LEASAM$^+$(SYSSAM$(f_{in}, start, size), Sim, \delta, f_{out})$;
      if $m - |f_{out}| < 0$
        then $progress := progress + 1$;
          else $\delta := update(adapt, start, size)$;
      endif
      $start := start + 1$;
  enddo
  if $|f_{out}| \leq m$
    then return(f_{out});
      else return$(reduce(f_{out}, m))$;
  endif
end;

Within each sampling loop, SIMSAM selects leader tuples from systematic samples of a pre-specified size beginning at the current start position. Note, leader sampling still uses the current focusing output to seek for most similar existing prototypes in order to decide whether to add tuples from the current working set to the final focusing output. Thus, LEASAM$^+$ has an extra parameter which refers to the current focusing output. If the number of tuples in the current focusing output increases within this sampling loop, the algorithm increments the progress counter. Otherwise, if SIMSAM does not select additional prototypes, it automatically updates similarity threshold δ. Afterwards, the next loop begins at an increased start position for systematic sampling. Thereby, SIMSAM considers new candidates for selection of leaders within each loop.

As long as SIMSAM has not considered the entire focusing input (indicated by smaller values for *start* than for *step*), it repeats sampling, unless sampling does not make any progress in terms of selecting additional prototypes for a number of iterations (controlled by comparisons between the difference of the current *start* position and the current *progress* counter and the pre-specified number of *iterations*). In this case, SIMSAM stops sampling before it processed all tuples in the focusing input.

At the end, SIMSAM checks the size of the final focusing output. If this size exceeds a pre-specified number of tuples m, the algorithm invokes the *reduce* procedure to decrease the sample size to the desired number of tuples (see section 4.2.3.3, page 146ff). Otherwise, SIMSAM returns the final focusing output. Note, if we initially set sample size m to the number of tuples in the focusing input, we force SIMSAM to select as many prototypes as it regards as necessary.

Advantages and Disadvantages

In terms of the unifying framework of focusing solutions, similarity-driven sampling applies systematic sampling to select intermediate working sets in the first step. Then, SIMSAM uses a compound clustering and prototype selection procedure in order to decide which tuples in the current working set are added to the current focusing output. Similarity-driven sampling utilizes a simple evaluation criterion in terms of the number of iterations without progress in order to examine, if focusing stops or continues with selection of further tuples into the focusing output.

All in all, similarity-driven sampling further contributes to enhancements to existing efforts towards focusing solutions and overcomes additional deficits of pure leader sampling. The most important advantage of SIMSAM in comparison to LEASAM is that similarity-driven sampling no longer depends on fixed similarity thresholds and does not necessarily process the entire focusing input, if subsets probably lead to sufficient focusing outputs.

Now, we describe automatic estimation and adaptation of similarity thresholds in similarity-driven sampling as well as procedure *reduce* in SIMSAM in more detail.

4.2.3.2 Estimation and Adaptation of Similarity Thresholds

Focusing success of leader sampling mainly depends on an appropriate choice of similarity threshold δ, and the appropriateness of δ depends on the focusing

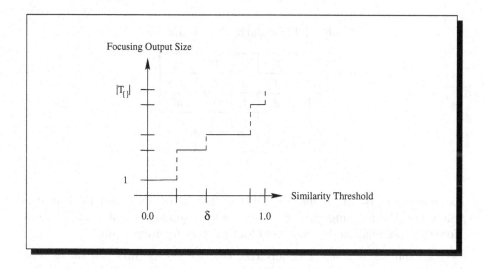

Figure 4.5: Similarity Thresholds and Focusing Output Size in LEASAM

context. For example, if the focusing input is homogeneous, higher values for similarity threshold δ are more appropriate. In contrast, high values for δ result in large samples, if the focusing input is heterogeneous. Although the number of tuples in the focusing output is not the only criterion to measure focusing success, focusing solutions aim at selection of a *reasonable* number of tuples. Consequently, the first general idea of appropriate similarity thresholds is to set δ to values such that leader samples from the focusing input contain a *reasonable* number of tuples. Thus, we first consider relations between similarity threshold values and resulting numbers of tuples in the focusing output, if we apply leader sampling.

Similarity Thresholds and Focusing Output Size

In general, the relation between similarity threshold δ and the number of tuples in the focusing output pursues the following tendency. The higher the similarity threshold value is, the higher is the number of tuples in the final focusing output. Vice versa, low threshold values result in smaller focusing outputs. For the two extreme thresholds, 0 and 1, LEASAM selects only the first tuple for $\delta = 0$ and removes exactly all duplicates in the focusing input for $\delta = 1$ (see proposition 2.5.1, page 41).

The number of different similarity threshold values that result in different numbers of tuples in the focusing output is finite. Figure 4.5 depicts the typical relation between increasing similarity thresholds and resulting focusing output sizes in leader sampling. For a fixed order of tuples in the focusing input, the number of tuples in the focusing output increases by a number of tuples at specific threshold values. In between two of these values, the focusing output

Table 4.2: Similarity Matrix Example

	t_1	t_2	t_3	t_4
t_1	1	δ	$< \delta$	$< \delta$
t_2	δ	1	δ'	δ'
t_3	$< \delta$	δ'	1	$< \delta$
t_4	$< \delta$	δ'	$< \delta$	1

size remains constant. Thus, the typical relation between similarity threshold values and the focusing output size in leader sampling is a step-like function. However, this relation does not hold for arbitrary focusing inputs.

For example, it is not generally true that increasing threshold values result in an increasing number of tuples in the focusing output. Table 4.2 outlines a counter example. If we assume $\delta < \delta'$, the application of leader sampling to $\{t_1, t_2, t_3, t_4\}$ generates a smaller focusing output for δ' ($\{t_1, t_2\}$) than for δ ($\{t_1, t_3, t_4\}$), although δ is smaller than δ'.

If we assume that we explicitly know the relation between different similarity threshold values and the resulting number of tuples in the focusing output for specific focusing inputs, we are able to analyze this relation in order to set the similarity threshold to an appropriate value. For example, if we specify a required sample size in advance, we compute the corresponding threshold value according to the given relation between similarity threshold values and focusing output sizes. Alternatively, we hill-climb the function, which depicts the relation between similarity threshold values and focusing output sizes, and select the similarity threshold below the maximum gap in focusing output sizes.

Costs for such analyses increase, if the number of different threshold values that result in different numbers of tuples in the focusing output is high. The number of different threshold values that result in different numbers of tuples in the focusing output depends on the number of different similarities between tuples in the focusing input. The more different similarities between tuples in the focusing input occur, the higher is the number of different threshold values that result in different numbers of tuples in the focusing output. If we apply the weighted cumulated similarity measure (see definition 2.5.9, page 40), the relation between similarity threshold values and the number of tuples in the focusing output becomes more and more continuous, if the relative number of quantitative attributes in comparison to the number of qualitative attributes increases.

The following considerations imply that the number of different similarities between tuples in the focusing input, and hence the number of different similarity threshold values that result in different numbers of tuples in the focusing output, is usually high.

Proposition 4.2.6 (Number of Local Similarities)

Assume a (database) table (A, T), and an attribute $A_j \in A$. We denote the number of different local similarities $sim_j(a_{jk}, a_{jk'})$, $a_{jk}, a_{jk'} \in dom(a_j)$ as $| sim_j |$. Then,

$$| sim_j | \leq \begin{cases} 2, & \text{if } a_j \text{ qualitative} \\ \dfrac{N_j \cdot (N_j - 1)}{2}, & \text{if } a_j \text{ quantitative} \end{cases}$$

Proof:

This follows from definitions 2.5.7 and 2.5.8. $\square$

Proposition 4.2.6 gives an upper bound for the number of different local similarities between different values of an attribute domain. In case of qualitative attributes, local similarity only distinguishes two cases. If two values are the same, local similarity is 1. Otherwise, local similarity is 0. Thus, we have at most two different local similarities, if an attribute is qualitative. In case of quantitative attributes, local similarities reveal at most as many different values as the attribute domain contains different pairs of values. The true upper bound for the number of different local similarities for quantitative attribute domains is usually smaller, since distances between two different pairs of attribute values can be the same.

Proposition 4.2.7 uses the number of different local similarities to compute an upper bound for the number of different weighted cumulated similarities between tuples in the focusing input. The upper bound for the number of different similarities between tuples in the focusing input equals the product of all numbers of different local similarities. The true upper bound for the number of different similarities is smaller, if relevance weights of two different attributes are the same, or the product of attribute relevance and local similarity is identical for two different attributes. For example, if all attributes are qualitative and all attribute relevance weights are different, the upper bound for the number of different similarities is 2^N.

Proposition 4.2.7 (Number of Similarities)

Assume a (database) table (A, T), and a focusing input $f_{in} \subseteq T$. We denote the number of different similarities $Sim(t_i, t_{i'})$, $t_i, t_{i'} \in f_{in}$ as $| Sim |$. Then,

$$| Sim | \leq \prod_{j=1}^{N} | sim_j |.$$

Proof:

This follows from definition 2.5.9. $\square$

Since the number of attributes in focusing contexts is usually high, the overall number of different similarities between tuples in the focusing input, and hence the number of different similarity threshold values that potentially result in different numbers of tuples in the focusing output in leader sampling, is also high. If we apply stratification before similarity-driven sampling and use SIMSAM within each stratum separately, we reduce the number of different similarities between tuples within each stratum since stratification reduces the number of tuples to consider in leader sampling, and hence the number of occurring attribute values also decreases. However, we probably still get an overall high number of different similarities.

In summary, it is computationally expensive to analyze the relation between similarity threshold values and the resulting number of tuples in the focusing output. Consequently, an exhaustive search among all similarity threshold values and their resulting focusing outputs in leader sampling for the best setting of δ is not feasible. Thus, we need to develop alternative heuristics to estimate appropriate similarity threshold values.

Definition 4.2.7 (Mucha's Similarity Threshold Estimation)

Assume a (database) table (A, T), a focusing input $f_{in} \subseteq T$, attribute relevance weights $w_j \geq 0$, $1 \leq j \leq N$, an expected focusing output size $m \geq 0$, and $\gamma \geq 0$. Mucha (1992), page 121, suggests the following estimation of δ:

$$\delta := 1 - \frac{\gamma \cdot \sum_{j=1}^{N} w_j \cdot \sigma_j^2(f_{in})}{\sqrt{m}}.$$

Definition 4.2.7 cites an alternative estimation for similarity threshold δ which is based on a constant factor γ, attribute relevance weights w_j, standard deviations σ_j in attribute domains, and an expected sample size m (Mucha, 1992). Without any experimental experience, we do not know, if this estimation is appropriate. However, computational efforts for this estimation are high since we have to compute all standard deviations explicitly. Beyond an additional constant parameter, it is also hard to specify an appropriate sample size in advance. In conclusion, we propose to develop simple heuristics that are both computationally cheap and independent of the unknown sample size.

Similarity Thresholds and Hierarchical Clustering

In order to develop appropriate heuristics, we now relate different similarity threshold values and their resulting focusing outputs to hierarchical clustering. For example, we assume a sequence of similarity threshold values $\delta < \delta' < \delta''$ and the focusing input illustrated in figure 4.6 a) (black and white bullets). If we apply LEASAM$^+$ with the smallest similarity threshold value δ and assume that tuple number 1 is the first tuple in the focusing input, leader sampling

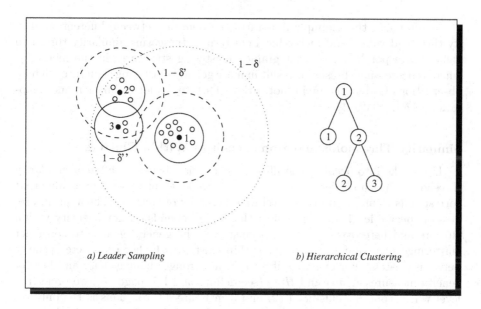

Figure 4.6: Leader Sampling and Hierarchical Clustering

selects only the first tuple as the focusing output. This prototype represents all tuples within the dotted circle with radius $1 - \delta$ (see figure 4.6 a)).[13] In terms of hierarchical clustering, this prototype represents the root cluster that contains all tuples in the focusing input (see figure 4.6 b)).

For the second similarity threshold value δ', the first tuple no longer covers all tuples but only represents tuples within the dashed circle with radius $1 - \delta'$ (see figure 4.6 a)). Thus, leader sampling selects an additional prototype to represent tuples outside this area. We assume, leader sampling encounters tuple number 2 as the next prototype. Then, the focusing output contains two tuples that represent all tuples in the focusing input which is now separated into two clusters. The resulting clustering structure refines the initial clustering. Leader sampling with similarity threshold δ' separates the root cluster into two smaller subsets of tuples denoted by number 1 and 2 (see figure 4.6 b)).

If we now increase the similarity threshold again, leader sampling produces a third focusing output which corresponds to a more fine-grained clustering structure again. For δ'', LEASAM selects three prototypes to represent all tuples in the focusing input. We assume, tuple number 3 is the next prototype. Full line circles with radius $1 - \delta''$ indicate the subset of tuples represented by each selected prototype (see figure 4.6 a)). In terms of hierarchical clustering, the focusing output represents clusters at the next lower level in the hierarchy. The second cluster of the previous clustering is now separated into two clusters, number 2 and 3 (see figure 4.6 b)).

---

[13]Note, in this example we assume an euclidian variant of weighted cumulated similarity.

In summary, this example illustrates the relation between different similarity threshold values and hierarchical clustering. Increasing similarity threshold values correspond to more fine-grained clustering structures in the hierarchy, whereas decreasing thresholds result in more general clustering structures. These observations lead to the main motivation of strategies for estimation and adaptation of δ in SIMSAM.

Similarity Thresholds and Homogeneities Within Clusters

If we take into account the illustrated relation between different similarity threshold values and hierarchical clustering, we attempt to set δ to a value that corresponds to an appropriate level in a hierarchy of clusters. The appropriateness of specific levels in hierarchical clustering depends on homogeneity within clusters and heterogeneity between clusters. The general goal is to construct clusterings such that homogeneity within clusters is higher than heterogeneity between clusters. For example, if we use the stronger homogeneity and heterogeneity measures Hom_2 and Het_2 (see definition 4.1.4, page 96), we prefer the level which contains clusters 1, 2, and 3 in figure 4.6 b). This is the only set of clusters with higher homogeneity within clusters than heterogeneity between clusters.

In order to develop efficient heuristics for estimations of appropriate similarity threshold values that correspond to appropriate levels in hierarchical clustering, we now analyze the relation between δ and homogeneity within clusters. Proposition 4.2.8 states that the lower bound for minimum similarities within subsets of tuples which are represented by the same prototype directly depends on the similarity threshold value in leader sampling. Thus, δ does not only control selection of prototypes but also regulates homogeneity within subsets of tuples which are represented by the same prototype. For high similarity threshold values, we automatically get high homogeneity within subsets of tuples which are represented by the same prototype.

Proposition 4.2.8 (Homogeneity within Leader Clusters)

Assume a (database) table (A, T), a focusing input $f_{in} \subseteq T$, $0 \leq \delta \leq 1$, a focusing output $f_{out} = \text{LEASAM}(f_{in}, Sim, \delta) \subseteq f_{in}$, and $T(t_i) := \{t_{i'} \in f_{in} \mid t_i \text{ is leader of } t_{i'}\}$. Then,

(i) $\forall t_i \in f_{out} : Hom_1(T(t_i)) \geq 2 \cdot \delta - 1$, and

(ii) $\forall t_i \in f_{out} : Hom_2(T(t_i)) \geq 2 \cdot \delta - 1$.

Proof:

(i) LEASAM (see algorithm 4.2.2, page 116, and definition 4.2.1, page 116) represents an arbitrary tuple $t_{i'} \in f_{in}$ by $t_i \in f_{out}$ if $Sim(t_{i'}, t_i) \geq \delta$. Thus, $Sim(t_{i'}, t_i) \geq \delta$ for all leaders $t_i \in f_{out}$ and for all tuples $t_{i'} \in T(t_i)$.

As a consequence of proposition 2.5.2, we also have $Sim(t_{i'}, t_{i''}) \geq Sim(t_{i'}, t_i) + Sim(t_i, t_{i''}) - 1$ for all leaders $t_i \in f_{out}$ and for all tuples $t_{i'}, t_{i''} \in T(t_i)$. If we put both observations together, we get

$$
\begin{aligned}
Sim(t_{i'}, t_{i''}) &\geq Sim(t_{i'}, t_i) + Sim(t_i, t_{i''}) - 1 \\
&\geq \delta + \delta - 1 \\
&= 2 \cdot \delta - 1.
\end{aligned}
$$

This means by definition 4.1.4, $\forall t_i \in f_{out} : Hom_1(T(t_i)) \geq 2 \cdot \delta - 1$.

(ii) The second relation is a consequence of the first. $\qquad\square$

High homogeneity within subsets of tuples which are represented by the same prototype is advantageous. If we are not able to ensure high homogeneity within these subsets, i.e., if heterogeneous subsets of tuples are represented by the same prototype, leader sampling probably by-passes important information since heterogeneous subsets of tuples are likely to contain more information than accomplished by single prototypes. Consequently, high similarity threshold values are preferable.

Similarity Thresholds in SIMSAM

In general, it is advisory to adapt similarity threshold values dynamically during similarity-driven sampling. If we fix the similarity threshold value, SIM-SAM only draws representatives at a single level in the corresponding hierarchical clustering. If we vary the similarity threshold value, SIMSAM is able to select representatives at different levels of granularity and to seek for the most appropriate value that corresponds to the most appropriate level in hierarchical clustering. Moreover, if we vary the similarity threshold, SIMSAM is also able to keep representatives of some clusters at higher levels, whereas it moves down to more specific levels for other clusters if necessary. Thus, dynamic adaptation of similarity threshold values during similarity-driven sampling is beneficial.

In the following, we assume the existence of a *true* hierarchical clustering with *true* homogeneities within subsets of sufficiently similar tuples at different levels of granularity. Since we aim at selection of a single prototype within each cluster at the most appropriate level, we want to set similarity threshold δ close to the value that corresponds to the true homogeneity within clusters at the most appropriate level. If the homogeneity within clusters at this level is high, we also attempt to set the similarity threshold to a high value. Similarly, if the homogeneity within clusters at the most appropriate level is low, we propose to use low similarity threshold values.

Since occurring similarities between tuples and their representative prototypes yield estimates for true homogeneity within clusters, it is a reasonable strategy to adapt the similarity threshold value according to observed similarity values in leader sampling. For this reason, we keep minimum, average, and maximum similarities between tuples and their representatives as aggregated information on occurring similarities in similarity-driven sampling. In order to avoid extreme situations, we ignore occurring similarities of 0 and 1. Since we keep similarity characteristics across different sets of represented tuples, we are able to estimate the overall clustering quality at the corresponding level in hierarchical clustering. The more similarity comparisons between tuples and leaders we take into account, the more reliable is the estimation of corresponding homogeneities within clusters.

Procedure 4.2.4   *update(adapt,start,size)*

---

begin
    if *adapt* = *min*
      then $\delta := \min\limits_{i=1}^{start \cdot size} \{sim_i\};$
    endif
    if *adapt* = *avg*
      then $\delta := \dfrac{1}{start \cdot size} \cdot \sum\limits_{i=1}^{start \cdot size} sim_i;$
    endif
    if *adapt* = *max*
      then $\delta := \max\limits_{i=1}^{start \cdot size} \{sim_i\};$
    endif
    return(δ);
end;

Since SIMSAM retains different similarity characteristics (minimum, average, and maximum), it is able to utilize alternative estimation policies. If SIMSAM uses the *minimum strategy* and adapts δ to the minimum of observed similarities, homogeneity within corresponding clusters is low, but heterogeneity between corresponding clusters is high. In this case, similarity-driven sampling only selects a few prototypes within a few clusters. If SIMSAM applies the *maximum strategy*, homogeneity within corresponding clusters is high, but heterogeneity between corresponding clusters is low. Then, similarity-driven sampling selects more prototypes that represent smaller homogeneous clusters. The *average strategy* is in between minimum and maximum strategies, and results in a compromise between homogeneity within clusters and heterogeneity between clusters.

Procedure 4.2.4 summarizes alternative adaptation strategies in similarity-driven sampling. This procedure assumes that SIMSAM keeps sequences of observed similarities sim_i between the ith tuple in the focusing input and its most similar leader in the current focusing output. The true implementation of *update* in similarity-driven sampling does not store all similarities but updates characteristic similarity values immediately after similarity comparisons. According to the selected adaptation strategy, procedure *update* returns the minimum, average, or maximum observed similarity. The product of *start* and *size* corresponds to the overall number of tuples in the focusing input that SIMSAM processed before this adaptation step. The *start* value indicates the number of systematic samples in SIMSAM up to this point, and *size* refers to the number of tuples within each of these systematic samples.

Note, procedure *update* includes possibilities to increase similarity threshold values as well as to decrease δ. Thereby, SIMSAM moves up and down in the corresponding hierarchical clustering. Since adaptation of similarity thresholds in SIMSAM sets the current threshold to the current estimation of the true homogeneity within clusters of the corresponding clustering hierarchy, similarity threshold δ and the estimated homogeneity within clusters always coincide after an adaptation step in similarity-driven sampling.

Note also, we only tune similarity threshold values with respect to homogeneity within clusters and do not explicitly consider heterogeneity between clusters since we regard homogeneity within clusters as more important than heterogeneity between clusters for focusing purposes.

In order to decide whether SIMSAM needs to update the current similarity threshold, it considers sampling progress on the current working set. If SIMSAM does not make progress, i.e., if it does not select additional prototypes, the current similarity threshold is either an inappropriate estimation of homogeneity within clusters, or it already matches *true* homogeneity, and it is not necessary to select more tuples in order to represent information in the focusing input at the most appropriate level in the corresponding hierarchical clustering. For distinction between these two cases, we use the following strategy.

If SIMSAM does not make progress a *few* times, we adapt similarity threshold δ in order to give similarity-driven sampling a chance to make progress at different levels of granularity in the corresponding hierarchical clustering. If SIMSAM does not make progress for a pre-specified number of iterations, we assume that the current similarity threshold value correctly estimates *true* homogeneities within the focusing input, and that SIMSAM does not need to select more tuples to represent the entire focusing input.

At this point, we notice that the usage of systematic sampling to select intermediate working sets from the focusing input with varying start positions is particularly advantageous, if we use sorting according to order relation $\succ$ before the application of similarity-driven sampling. Since similar tuples are likely to have close positions in sorted focusing inputs, each working set is likely to represent the whole range of information in the focusing input. Thus, SIMSAM approximately considers the whole information in the focusing input at each level of the clustering hierarchy which corresponds to the current value of similarity threshold δ.

4.2.3.3   Remove Less Representative Prototypes

Prototypes represent subsets of sufficiently similar tuples at different levels of representativeness. In similarity-driven sampling, we introduce a simple measure of representativeness. We count the number of tuples in the focusing input that each prototype in the focusing output represents and consider this number as the level of representativeness of prototypes (see definition 4.2.8). Each time SIMSAM retrieves a prototype in the current focusing output as the nearest neighbor of the current tuple in the focusing input, it increments weight w_i of this prototype. If more than a single prototype have the same highest similarity to the current tuple, similarity-driven sampling only increments the weight of the *first* nearest neighbor in the current focusing output. The initial value of each prototype weight is 1, since at the time SIMSAM adds new prototypes to the focusing output, prototypes only represent themselves.

Definition 4.2.8 (Prototype Weights)

Assume a (database) table (A, T), a focusing input $f_{in} \subseteq T$, and a focusing output $f_{out} = \text{SIMSAM}(f_{in}, m, Sim, adapt, size, iterations) \subseteq f_{in}$. We define the prototype weight of $t_i \in f_{out}$ as

$$w_i := | \{t_{i'} \in f_{in} \mid t_i \text{ is the first nearest neighbor of } t_{i'}\} |.$$

At the end of focusing in similarity-driven sampling, prototype weights reflect levels of representativeness for each prototype. If prototypes reveal higher weights than other prototypes, these prototypes are at higher levels of representativeness than the others since they represent more tuples. For this reason, it is necessary to retrieve the most similar prototype instead of the first tuple in the current focusing output which exceeds similarity threshold δ in comparison to the current tuple in the focusing input. If SIMSAM only retrieves the first prototype which exceeds similarity threshold δ, early selected prototypes have higher probabilities to get higher prototype weights. Alternatively, SIMSAM may also increment prototype weights of each prototype which exceeds the similarity threshold during retrieval. Then, tuples in the focusing input are eventually represented by more than single prototypes in the focusing output.

Procedure 4.2.5 characterizes the usage of prototype weights in order to remove less representative prototypes from the focusing output in similarity-driven sampling. The general idea is to prefer prototypes that represent many tuples since these prototypes represent more important information in the focusing input. If the final focusing output contains more tuples than a pre-specified focusing output size, SIMSAM starts removing prototypes that only represent themselves.[14] If the focusing output still contains more tuples than desired after

---

[14]Note, deletion of tuples shifts the last tuple to the position of removed tuples. Hence, SIMSAM only increments i in procedure *reduce*, if it does not remove a prototype.

removal of prototypes at the lowest level of representativeness, SIMSAM increments prototype threshold ϵ and proceeds elimination of prototypes according
to the new minimum number of represented tuples required for prototypes at an
increased level of representativeness. In the next iteration of *reduce*, SIMSAM removes prototypes that only represent a single additional tuple. Similarity-driven
sampling iterates this process until the focusing output does not contain more
than the pre-specified number of tuples.

Procedure 4.2.5   *reduce*(f_{out}, m)

---

begin
   $i := 1$;
   $\epsilon := 1$;
   while $| f_{out} | > m$ **do**
        while $i \leq | f_{out} |$ **do**
            if $w_i \leq \epsilon$
               then $f_{out} := f_{out} - \{t_i\}$;
               else $i := i + 1$;
            endif
        enddo
        $i := 1$;
        $\epsilon := \epsilon + 1$;
   enddo
   return(f_{out});
end;

Note, procedure 4.2.5 uses pre-specified sample sizes to control the value of
prototype threshold ϵ. Alternatively, it is also possible to specify fixed prototype thresholds and to remove all prototypes that represent less than ϵ tuples,
regardless the resulting sample size. However, it is usually difficult to set ϵ to an
appropriate value in advance, and it is also more likely that we know the number
of tuples that the following data mining algorithm is able to handle rather than
the most appropriate level of representativeness.

Prototype Weights and Outliers

Although the primary purpose of using procedure 4.2.5 in SIMSAM is to keep
control of the final focusing output size, removing less representative prototypes
roughly corresponds to removing *outliers*. On the one hand, outliers often match
noisy tuples. Then, removing outliers as noise is often useful since data mining
algorithms usually yield less appropriate results, if they learn on noisy tuples. On
the other hand, outliers possibly represent rare information in the focusing input.
If only a single outlier represents this particular information, it is usually more
appropriate to retain this outlier in order to avoid to neglect its information.

Since we suggest to stratify the focusing input before SIMSAM, we automatically avoid the second case when removing less representative prototypes in similarity-driven sampling. If tuples represent outlying interesting information, stratification generates an extra stratum for this information. Then, these tuples are definitely put into the focusing output, since the minimum sample size within each stratum is 1. Thus, SIMSAM selects outlying tuples as the first prototype within their stratum and does not remove them due to the minimum sample size requirement within each stratum.

Further Usage of Prototype Weights

We expect that prototype weights provide additional advantages for data mining in the selected focusing contexts. Since top down induction of decision trees mainly uses relative frequencies to compute information gain ratio in order to select the best split attribute, the weighting mechanism approximately represents original frequencies in the focusing input. If we apply nearest neighbor classifiers without any voting procedure, prototype weighting does not make any difference.

We also expect that the maximum estimation and adaptation strategy to control similarity threshold values in combination with elimination of less representative prototypes is especially appropriate. The first strategy initially selects more prototypes than probably necessary, whereas the second mechanism restricts the resulting focusing output to prototypes at higher levels of representativeness.

4.2.3.4   Attribute Relevance Weights

Appropriate estimations of attribute relevance are important in more intelligent sampling at several points. First, the weighted cumulated similarity measure (see definition 2.5.9, page 40) uses attribute relevance weights to reflect the contribution of each attribute to cumulated similarity between tuples, and we use similarity measures in advanced leader sampling as well as in similarity-driven sampling. Second, sorting and stratification (see definition 4.2.2, page 117, and section 4.2.2.4, page 120ff, respectively) consider attributes in order of decreasing relevance. Both procedures take into account the most important attribute first, and then iteratively proceed with the next important attribute, if the termination criterion does not hold yet.

A straightforward approach to estimate attribute relevance in KDD projects is to ask domain experts about their experience and their assessment of attribute relevance in the specific context of their project. However, opinions of domain experts only represent specific personal preferences, and attribute relevance probably differs from expert to expert. Thus, estimations of attribute relevance according to domain experts are largely subject to personal preferences and often do not assess attribute relevance appropriately. Hence, we argue for the development of automatic estimations of attribute relevance.

Information Gain Ratio Attribute Relevance Weights

Since we are dealing with attribute relevance in specific focusing contexts, we suggest to take into account data mining goal, data characteristics, and data mining algorithm. For example, in case of classification and top down induction of decision trees, we propose to utilize the same information theoretic approach as the data mining algorithm. Thus, we compute attribute relevance according to information gain ratio as if we separate the entire focusing input using the respective attribute as the split attribute in top down induction of decision trees. Definition 4.2.9 states this particular approach precisely.

Definition 4.2.9 (Attribute Relevance Weights)

Assume a (database) table (A, T), an attribute $a_j \in A$, a class attribute $a_N \in A$, and a focusing input $f_{in} \subseteq T$. We define relevance weight w_j of attribute a_j as

$$w_j := I_j^+(f_{in}).$$

Alternative Attribute Relevance Weights

If the focusing context is different, we suggest to apply different attribute relevance weights. Table 4.3 depicts example contexts and alternative functions to automatically compute attribute relevance as well as references for more detailed definitions and descriptions of these approaches. For example, we conducted experiments with APRIORI (Agrawal *et al.*, 1993) to identify association rules on data of car histories at Daimler-Benz, and it turned out that we achieve the best results, if we use the same measure to compute attribute relevance as APRIORI applies to select associations.[15] Note, implementations of more intelligent sampling techniques supply all measures listed in table 4.3.

For overviews of further alternative functions to automatically compute attribute relevance in machine learning, we refer to Wettschereck (1994), Wettschereck *et al.* (1995), Kohavi *et al.* (1997), Blum & Langley (1997).

4.3 A Unified Approach to Focusing Solutions

In summary, enhancements to the unifying framework of existing state of the art approaches towards focusing solutions in more intelligent sampling techniques contain the following novel mechanisms:

---

[15]These results are not reported in detail here.

Table 4.3: Focusing Contexts and Attribute Relevance

Focusing Context	Attribute Relevance	Reference
Decision Trees	Symmetric Information Gain Ratio	Mantaras, 1991
CART	Gini Index	Breiman et al., 1984
CART	Normalized Gini Index	Zhou & Dillon, 1991
	Relevance	Baim, 1988
APRIORI	χ^2-Measure	Borgelt et al., 1996
K2	$log_2(g)/N$	Cooper & Herskovits, 1992
MDL	Reduction of Description Length	Krichevsky & Trofimov, 1983; Rissanen, 1987
Possibilistic Networks	Specificity Gain	Higashi & Klir, 1982; Klir & Mariano, 1987; Borgelt et al., 1996

- *Sorting*

- *Stratification*

- *Leader Sampling*

- *Similarity-driven Sampling*

These enhancements include various auxiliary procedures such as alternative methods for automatic computation of attribute relevance and several discretization approaches. We propose to integrate all enhancements into a unified approach to focusing solutions. *Generic sampling* implements this unified approach and allows specifications of numerous different focusing solutions. Each specific parameter setting in generic sampling corresponds to a single concrete focusing solution. In this section, we describe the generic sampling approach and its implementation in a commercial data mining system.

4.3.1   Generic Sampling

Since the appropriateness of focusing solutions depends on the focusing task and its focusing context, we unify enhancements to the unifying framework of focusing solutions in generic sampling in order to supply different focusing solutions. Figure 4.7 outlines the unified approach to focusing solutions in generic sampling which incorporates the following three steps:

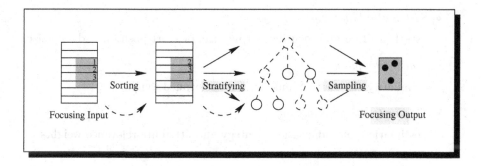

Figure 4.7: Generic Sampling

1. *Sorting*

2. *Stratifying*

3. *Sampling*

Instantiations of generic sampling select combinations of sorting, stratification, and sampling to generate specific focusing solutions. Whereas the optional first two steps prepare the original focusing input, applications of sampling techniques select final focusing outputs.

The first step in generic sampling sorts the original focusing input according to order relation $\succ$ using an implementation of quick sort. Sorting depends on attribute relevance weights, and generic sampling distinguishes between manual specification and automatic computation of relevance weights. It is also possible to skip the first preparation step and to immediately proceed with the next step. In generic sampling, the next step is stratification of the original or sorted focusing input. Beyond attribute relevance weights, stratification additionally requires selection of a discretization approach, if the data characteristics contain quantitative attributes, and the definition of a maximum stratum size. The result of the second preparation step is a set of strata, and each stratum contains a subset of similar tuples. Again, it is also possible to pass the original or sorted focusing input to the next step in generic sampling without stratification. If we skip stratification, generic sampling proceeds as if stratification generates only a single stratum. Finally, the sampling step applies a specific sampling technique to each stratum and unifies the resulting samples to the final focusing output.

In generic sampling, we provide four different sampling techniques with the following parameters:

- *Simple Random Sampling with Replacement*
  - focusing output size

- *Systematic Sampling*
  - start position and focusing output size, or start position and step size
- *Leader Sampling*
  - similarity threshold δ and attribute relevance weights
- *Similarity-driven Sampling*
  - estimation and adaptation strategy and attribute relevance weights

---

Algorithm 4.3.1 $\textsc{GenSam}(f_{in}, m, relevance, max, discretize, \textsc{Sampling})$

---

begin
  $f_{out} := \emptyset$;
  if *relevance?*
    then $W := relevance(f_{in})$;
  endif
  if *sort?*
    then $f_{in} := \textsc{Sort}(f_{in}, \succ)$;
    else $f_{in} := \textsc{Shuffle}(f_{in})$;
  endif
  if *stratify?*
    then $S := \textsc{Stratify}(f_{in}, N, max, discretize)$;
    else $S := \{f_{in}\}$;
  endif
  $l := 1$;
  while $l \leq |\,S\,|$ **do**
      $m_l := \lfloor \dfrac{|\,s_l\,|}{|\,f_{in}\,|} \cdot m \rfloor + 1$;
      $f_{out} := f_{out} \cup \textsc{Sampling}(s_l, m_l, \dots)$;
      $l := l + 1$;
  enddo
  return(f_{out});
end;

Algorithm 4.3.1 shows an implementation of generic sampling. Note, the complete set of parameters depends on the specific instantiation of generic sampling. $\textsc{GenSam}$ starts focusing with initialization of the focusing output to an empty set. If the selected instantiation requires attribute relevance weights, $\textsc{GenSam}$ proceeds with specification of attribute relevance weights. It either reads these weights from an input file (e.g., specified by domain experts or resulting from previous runs) or automatically computes relevance weights using the selected *relevance* procedure. Then, generic sampling sorts the original focusing input according to order relation $\succ$, or $\textsc{GenSam}$ shuffles the original focusing input into a random order, if the specific instantiation skips the first preparation step. Similarly, $\textsc{GenSam}$ stratifies the resulting focusing input, if·

the specific instantiation desires stratification, or it initializes the set of strata to a single stratum that contains the entire focusing input.

For each of the resulting strata, generic sampling applies the specified SAM-PLING technique and unifies the resulting samples to the final focusing output. The proportion of each stratum in the entire focusing input determines the sample size within each stratum. In case of leader and similarity-driven sampling, it is also possible to automatically select focusing outputs without pre-specified focusing output sizes. Note, in case of leader sampling we also allow approximations of pre-specified focusing output sizes by the usage of SIMSAM's *reduce* procedure. At the end, GENSAM returns the overall focusing output.

Advantages and Disadvantages

In comparison to the unifying framework of existing focusing solutions, generic sampling extends the framework by two additional preparation steps, sorting and stratification, and keeps the ability to use statistical sampling techniques, random sampling with replacement and systematic sampling. Since stratification corresponds to clustering, GENSAM also allows construction of clusters and then applications of sampling techniques to each cluster. Generic sampling also inherently includes clustering and prototyping in form of two more intelligent sampling techniques, leader sampling and similarity-driven sampling. Both techniques unify clustering and prototyping into a compound step.

The key advantage of generic sampling is its flexibility to allow specifications of numerous different focusing solutions. Data mining engineers instantiate the unified approach to focusing solutions according to their own preferences and dependent on the current focusing context. For example, they use their experiences in data mining to select the most appropriate focusing solution in the current KDD project.

4.3.2 Generic Sampling in a Commercial Data Mining System

The flexibility to specify numerous focusing solutions in generic sampling is also a pitfall, if it is not possible to select specific instantiations easily and to integrate the resulting focusing solutions into KDD processes properly. In order to enable easy usage of generic sampling for all types of users and straightforward integrations of focusing solutions into KDD processes, we integrate GENSAM into a commercial data mining system. As the data mining system, we choose the CLEMENTINE system since CLEMENTINE provides an intuitive visual programming interface which permits comfortable applications of many alternative techniques in almost all phases of KDD processes (Khabaza & Shearer, 1995). The CLEMENTINE system is developed and distributed by *Integral Solutions Limited* (ISL), a leading data mining tool vendor in United Kingdom.

In the following, we first outline characteristics of CLEMENTINE and briefly

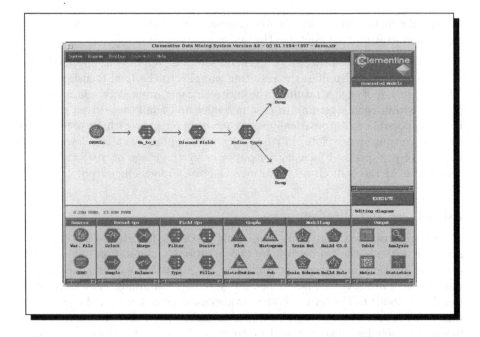

Figure 4.8: CLEMENTINE Data Mining System

describe the context of the CITRUS project which develops enhancements to CLEMENTINE beyond focusing solutions presented in this dissertation. Then, we characterize the integration of generic sampling into CLEMENTINE in the context of the CITRUS project.

CLEMENTINE

Figure 4.8 shows the CLEMENTINE data mining system. The upper part of the snapshot depicts the main menu, the workspace, and the model palette, whereas the lower part indicates six different categories of KDD tools for various phases in KDD processes. Each icon represents a single mechanism to access data sources, to operate on tuples (*records*) or attributes (*fields*), to visualize information on the data, to generate models, or to create outputs of results. CLEMENTINE users select icons from palettes at the bottom and connect them on the workspace above in order to build data flows of specific KDD processes. Each data flow is called a *stream*, and since icons are connected by arcs, these icons are also referred to as *nodes*. Each node provides its own *dialogue* which allows simple modifications of its parameters and alternative preferences.

Figure 4.8 shows a simple stream with data flow from left to right. This stream analyzes data on drugs and builds classifiers to predict the most appropriate drug for patients given some symptoms. The left hand node is a *variable file node* and accesses input data from a file. The second node is a *derive node*

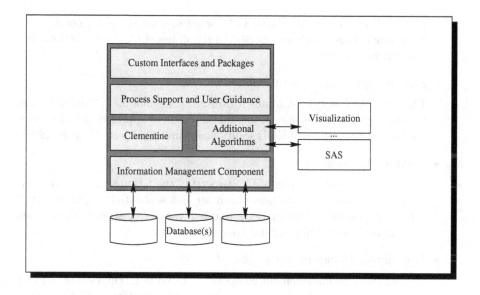

Figure 4.9: CITRUS Architecture

which generates an additional attribute for information on the relation between attributes NA and K. The third node is a *filter node* and discards irrelevant attributes. The following *type node* sets information on attribute domains and defines the class attribute for subsequent data mining algorithms. In this stream, we use neural nets at the top and top down induction of decision trees at the bottom to generate classifiers.

CITRUS

In the CITRUS project, we extend CLEMENTINE towards a process-oriented KDD tool (Wirth *et al.*, 1997). Daimler-Benz, ISL, and the University of Karlsruhe collaborate on this project. Figure 4.9 depicts the overall CITRUS architecture. The CITRUS architecture includes several additional components beyond CLEMENTINE as well as interfaces to external sources and tools. Currently, the CITRUS system supports access to ORACLE databases and provides interfaces to additional visualization programs as well as to the statistical package SAS.

The internals of the CITRUS architecture contain the following components:

- *Custom Interfaces and Packages*

  On top of the CITRUS architecture, we provide application specific customizations and packages which support instantiated KDD processes to solve concrete business questions. For example, we developed an *early indicator cars* prototype that supplies extra menus and extra dialogues to set all required parameters for the early indicator method (Wirth & Reinartz,

1996). With custom interfaces and packages, naive users do not need to know any details which are specific to the usage of CLEMENTINE or other components in CITRUS.

- *Process Support and User Guidance*

  The CITRUS prototype contains a user guidance module to support users in defining business questions, to guide selection of appropriate techniques, and to reuse sequences of KDD solutions in similar projects (Engels, 1996).

- *Additional Algorithms*

  In order to extend facilities of CLEMENTINE, the CITRUS architecture implements additional algorithms which are not supported in CLEMENTINE yet. For example, Daimler-Benz realized various clustering algorithms, association rule techniques, and bayesian networks.

- *Information Management Component*

  The information management component (IMC) in CITRUS enables direct access to relational databases, optimization of execution of CLEMENTINE streams, and intelligent materialization, maintenance, and reuse of data mining results (Breitner *et al.*, 1995). Therefore, the IMC builds an object-oriented schema of data as well as results and transforms original streams in CLEMENTINE into database queries for automatic execution of streams in standard SQL.

Generic Sampling in CLEMENTINE **and** CITRUS

Generic sampling is a particular enhancement to CLEMENTINE in the context of the CITRUS project. We implemented all components of generic sampling in C since C is an efficient programming language as well as widely available on different platforms. Since the implementation language of CLEMENTINE is *Poplog*, we integrated the C program into CLEMENTINE by providing an user interface to generic sampling written in Poplog. Generic sampling is now executable as an extra *focus node* on the record operations palette in CLEMENTINE, and CLEMENTINE users specify parameter settings in GENSAM in dialogues of focus nodes.

Figure 4.10 shows two examples of the user interface to generic sampling as a dialogue in CLEMENTINE. Figure 4.10 a) illustrates the dialogue of the focus node and its default settings in simple mode. This dialogue contains two tick boxes for sorting and stratification. If the user enables the sorting box, generic sampling sorts the original focusing input first. If the user enables the stratification box, focusing proceeds with stratification of the focusing input.

The sampling option of the focus node is a popup menu on mouse-click. It offers four different sampling techniques, and the user selects between *random*, *systematic, leader,* and *similarity-driven sampling* which correspond to the four available sampling techniques in generic sampling. In the size slot, the user

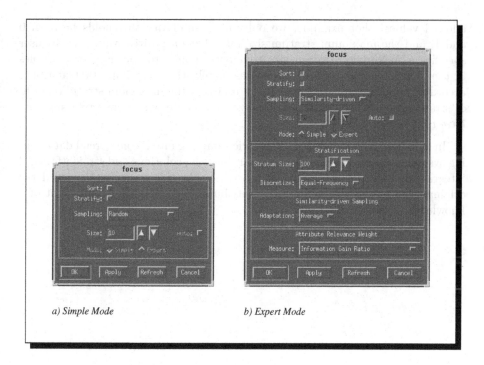

a) Simple Mode   *b) Expert Mode*

Figure 4.10: Generic Sampling in CLEMENTINE

specifies the required focusing output size in terms of the number of tuples or ticks the *auto* option. This option is not valid for simple random sampling and hence the focus node automatically disables the auto tick box, if random is the current selection in the dialogue. The focus node supplies automatic focusing output size selection for systematic sampling (i.e., the user indirectly specifies the focusing output size by start position and step size), leader sampling, and similarity-driven sampling.

Figure 4.10 b) gives an example of the focus node dialogue in expert mode. In this example, the user chooses an instantiation of generic sampling which uses sorting, stratification, similarity-driven sampling, and automatic selection of the focusing output size. Since the user enabled expert mode, the focus node dialogue now provides access to more details of parameters in generic sampling according to this choice of simple mode parameters. For example, the user additionally specifies a maximum stratum size of 100 tuples, selects equal-frequency discretization for stratification, applies the average estimation and adaptation strategy for similarity threshold δ, and chooses information gain ratio to compute attribute relevance. Note, the focus node automatically sets expert mode parameters to reasonable default values, if users apply focusing in simple mode.

In order to avoid that users specify invalid values for parameters in generic sampling, the focus node checks for each value whether it is in the range of allowed values. For example, we only allow similarity thresholds between 0 and 1, and the maximum stratum size must have a positive value. If the user specifies a maximum stratum size that corresponds to the number of tuples in the focusing input, stratification uses merely the most important attribute according to attribute relevance. If the user sets the maximum stratum size to 1, stratification considers all attributes except those which are irrelevant (i.e., have relevance weights 0).

In summary, the integration of generic sampling into a commercial data mining system supports easy usage of various focusing solutions and straightforward integrations of focusing into KDD processes. At this point, we achieved the original goal of this dissertation to contribute to the focusing component of a knowledge discovery system.

Chapter 5

Analytical Studies

In this chapter, we analyze the relation between focusing solutions and different focusing contexts. We present an average case analysis for specific classification goals and nearest neighbor classifiers. This analysis estimates the expected average classification accuracy on simple random samples in comparison to an ideal focusing solution. We experimentally validate the theoretical claims in some artificial domains. The primary goal of this chapter is an initial understanding of the appropriateness of focusing solutions in relation to focusing contexts.

5.1   An Average Case Analysis

In the previous chapters, we defined focusing tasks in detail, analyzed the state of the art of existing efforts towards focusing solutions, and developed enhancements to the resulting unifying framework of focusing solutions. All in all, we implemented a generic sampling approach that covers many different focusing solutions as instantiations. The main advantage of the large number of possible focusing solutions that result from parameter settings in GENSAM is the wide applicability of generic sampling in many different focusing contexts. Data mining engineers analyze the focusing context and specify most appropriate instantiations of generic sampling accordingly. The problem is to answer the question which particular focusing solution is best suited in a given focusing context.

This question motivates the contents of this chapter. An *average case analysis* helps to understand and characterize the relation between focusing tasks and focusing solutions and their focusing success. If we are able to predict focusing success of particular focusing solutions for specific focusing tasks, data mining engineers get an instrument to decide the appropriateness of focusing solutions in the concrete context of KDD projects. This question is also related to the question which focusing output size is necessary to achieve a pre-defined focusing success (see Kivinen & Mannila (1994), for example).

T. Reinartz: Focusing Solutions for Data Mining, LNAI 1623, pp. 159-172, 1999
© Springer-Verlag Berlin Heidelberg 1999

The general goal of an average case analysis is to investigate quality of data mining results for specific data mining algorithms in relation to the context of its application. For example, if we consider classification goals and measure quality of data mining results by classification accuracies, an average case analysis estimates the expected average classification accuracy in relation to data characteristics of the training set. In this section, we utilize the concept of an average case analysis to explore quality of data mining results, if we apply focusing solutions to select smaller training sets before the data mining phase. In order to validate theoretical claims of an average case analysis, we also run a set of experiments on artificial domains and compare experimental results with estimations of the analysis. Thereby, we combine advantages of theoretical approaches and experimental studies.

An average case analysis that covers the relation between focusing tasks and focusing solutions and their focusing success in general cases is not possible. An average case analysis is only feasible, if we make specific assumptions about the focusing context and the focusing solutions. For example, Hirschberg & Pazzani (1991) present an average case analysis of a k-CNF learning algorithm, Langley *et al.* (1992) consider bayesian classifiers for monotone conjunctive target concepts and independent, noise-free, boolean attributes, and Langley & Iba (1993) investigate nearest neighbor classifiers again with the same assumptions as for the analysis of bayesian classifiers plus a uniform distribution over the instance space. All of them perform average case analyses for special cases, and the complexity of their formulas already indicates obstacles of such theoretical approaches.

Well-Separated Clusterings

For these reasons, we also restrict the focusing contexts and focusing solutions which we consider for an average case analysis. In the following, we specify specific data characteristics which comprise a number of classification goals, select nearest neighbor classifiers as the data mining algorithm, and start the average case analysis with simple random sampling with replacement as the concrete focusing solution. As the criterion to measure focusing success, we focus our attention on quality of data mining results and use classification accuracies on the entire focusing input as the test set to rank the outcome of data mining. Definition 5.1.1 describes *well-separated clusterings*, and we assume this specific type of clusterings as the data characteristics to exemplify an average case analysis.

Definition 5.1.1 (Well-Separated Clustering (WSC))

Assume a (database) table (A, T), and an exhaustive, non-overlapping clustering C of T. C is a well-separated *clustering of f_{in}, if*

(i) $\forall c_l, c_{l'} \in C,\ l \neq l'$:

   $Het_2(c_l, c_{l'}) < Hom_2(c_l) \ \wedge \ Het_2(c_l, c_{l'}) < Hom_2(c_{l'})$,

(ii) $\forall c_l \in C \ \forall t_i, t_{i'} \in c_l$:

   $t_{iN} \equiv t_{i'N}$.

The data characteristics of well-separated clusterings contain two conditions (see definition 5.1.1). The first condition uses the definitions of homogeneity within clusters and heterogeneity between clusters (see definition 4.1.4, page 96). It requires that for all pairs of distinct clusters heterogeneity between clusters is smaller than homogeneity within each of these clusters. This means that maximum similarities between tuples of different clusters are always smaller than minimum similarities between tuples within each of these clusters. The first condition implies that the nearest neighbor of each tuple is always in the same cluster as the tuple itself. The second condition ensures that all tuples within each cluster belong to the same class and consequently have identical values for the class attribute.[1]

Example 5.1.1 (Well-Separated Clustering)

Figure 5.1 shows examples of focusing contexts in a two-dimensional euclidian space. Letters A, B, and C refer to class labels. Figure 5.1 a) represents a well-separated clustering. This focusing context contains three clusters that include six tuples with class label A, nine tuples with class label B, and four tuples with class label C, respectively. For each pair of distinct clusters, heterogeneity between clusters is smaller than homogeneity within each of these clusters, if we assume that we measure similarity with the weighted cumulated similarity measure Sim (see definition 2.5.9, page 40). All tuples within each of the three clusters also have the same class label. Hence, this focusing context meets all conditions of well-separated clusterings.

Figure 5.1 b) depicts an example which is not a well-separated clustering. This focusing context violates both requirements in definition 5.1.1. First, the similarity between the lower right tuple with class label C and the upper left tuple with class label A is higher than the similarity between the upper left tuple with class label A and the lower right tuple with class label A. Thus, maximum similarity between tuples in the left cluster and the lower right cluster is higher than minimum similarity between tuples within the lower right cluster. Moreover, the second condition of well-separated clusterings does not hold in this focusing context, since the left cluster contains a tuple with class label B as well as tuples with class label C.

---

[1]As in the previous chapters, we assume that the last attribute with index N is the class attribute.

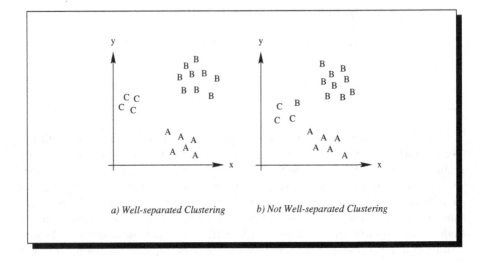

Figure 5.1: Well-Separated Clustering and Not Well-Separated Clustering

Note, although we restrict focusing contexts to well-separated clusterings, each arbitrary focusing input represents a well-separated clustering, if we consider each tuple as a single cluster and regard duplicates as a single tuple.

Accuracy of Nearest Neighbor Classifiers for WSCs

Now, we start the average case analysis for classification goals in domains which represent well-separated clusterings with an investigation of the quality of classification accuracies, if we apply nearest neighbor classifiers. The data characteristics of well-separated clusterings lead to close relations between the number of clusters represented in the training set and classification accuracy of nearest neighbor classifiers. The training set represents a cluster, if it contains at least a single tuple which is a member of this cluster. If the training set represents a cluster, the data characteristics of well-separated clusterings ensure that nearest neighbor classifiers return the correct class label for each tuple in the test set which belongs to the same cluster.

Proposition 5.1.1 (NN Accuracy for WSCs)

Assume a (database) table (A, T), a class attribute $a_N \in A$, a well-separated clustering C of T, a training set $T_{train} \subseteq T$, a nearest neighbor classifier NN which uses T_{train} and Sim to classify tuples, $C_R \subset C$, and $M_R := \sum_{l=1}^{|C_R|} |c_l|$. Then,

$$\forall c_l \in C_R \; \exists t_i \in T_{train} : t_i \in c_l \quad \Longrightarrow \quad a(NN, T) \geq \frac{M_R}{M}.$$

Proof:

Since $\forall c_l \in C_R \; \exists t_i \in T_{train} : t_i \in c_l$, and $\forall c_l, c_{l'} \in C$, $l \neq l'$: $Het_2(c_l, c_{l'}) < Hom_2(c_l) \; \wedge \; Het_2(c_l, c_{l'}) < Hom_2(c_{l'})$, as well as $\forall c_l \in C \; \forall t_i, t_{i'} \in c_l : t_{iN} \equiv t_{i'N}$, we also have $\forall c_l \in C_R \; \forall t_i \in c_l : NN(T_{train}, t_i, Sim) \equiv t_{iN}$. Since $M_R = \sum_{l=1}^{|C_R|} |c_l|$, this leads to $a(NN, T) \geq \dfrac{M_R}{M}$. $\qquad\square$

Proposition 5.1.1 states a lower bound for classification accuracy of nearest neighbor classifiers in relation to the number of represented clusters in the training set for focusing contexts which represent well-separated clusterings. We assume a training set which is a subset of all tuples in a (database) table and a test set which consists of all tuples. If a well-separated clustering of all tuples in the table exists and the training set represents a subset of all clusters, nearest neighbor classifiers return at least the correct class label for all tuples which belong to one of the represented clusters. Hence, classification accuracy of algorithm NN (see algorithm 2.5.1, page 37) is higher than or equal to the fraction of the number of tuples in represented clusters and the number of all tuples in the test set.

Note, proposition 5.1.1 also includes the following special case. If we assume that we use focusing outputs of focusing solutions as the training set and choose the entire focusing input as the test set, classification accuracy of nearest neighbor classifiers in focusing contexts which represent well-separated clusterings is 1, if a focusing solution selects a focusing output which contains at least a single tuple of each cluster. Thus, an *ideal* focusing solution in these focusing contexts is able to generate focusing outputs which represent all clusters in well-separated clusterings.

Although proposition 5.1.1 denotes a lower bound for the quality of data mining results in terms of classification accuracy of nearest neighbor classifiers, it does not take into account classification of tuples in non-represented clusters. In the following, we extend proposition 5.1.1 to an estimation of classification accuracies which also contains a component for non-represented clusters.

Whereas the previous result provides an exact lower bound for classification accuracies of nearest neighbor classifiers, the amount of correct classifications for tuples in non-represented clusters depends on the specific locations of clusters and relations between locations of represented and non-represented clusters. If the training set does not represent a cluster, nearest neighbor classifiers retrieve a tuple within a different cluster as the nearest neighbor of each tuple in non-represented clusters. Since the definition of well-separated clusterings only contains information on class labels within each cluster, we are not able to ensure that these nearest neighbors are members of the same class as the tuple which we want to classify. However, we possibly predict the correct class label for tuples in non-represented clusters in some cases.

Proposition 5.1.2 (NN Accuracy for WSCs II)

Assume a (database) table (A, T), a class attribute $a_N \in A$ with $dom(a_N) = \{a_{N1}, a_{N2}\}$, a well-separated clustering C of T, a training set $T_{train} \subseteq T$, a nearest neighbor classifier NN which uses T_{train} and Sim to classify tuples, $C_R \subset C$, $T_R := \bigcup_{l=1}^{|C_R|} c_l$, and $M_R := \sum_{l=1}^{|C_R|} | c_l | = | T_R |$. We define a function

$$\tilde{a} : \phi \times T^{\subseteq} \times T^{\subseteq} \longrightarrow [0; 1],$$

$$\tilde{a}(NN, T, T_R) := \frac{M_R}{M} + \left(1 - \frac{M_R}{M} \right) \cdot \left(\frac{n_{N1}(T_R)}{M} \cdot \frac{n_{N1}(T) - n_{N1}(T_R)}{M - n_{N1}(T_R)} \right.$$
$$\left. + \frac{n_{N2}(T_R)}{M} \cdot \frac{n_{N2}(T) - n_{N2}(T_R)}{M - n_{N2}(T_R)} \right).$$

Then,

$$\forall c_l \in C_R \; \exists t_i \in T_{train} : t_i \in c_l \quad \Longrightarrow \quad a(NN, T) \approx \tilde{a}(NN, T, T_R).$$

Proof:

We already know that nearest neighbor classifiers yield correct classifications for all M_R tuples in represented clusters in C_R (see proposition 5.1.1). This explains the first component in the definition of $\tilde{a}$.

For each of the remaining $M - M_R$ tuples, we estimate the probability of correct classification for each class label a_{N1} and a_{N2} separately. Each probability is the product of the probability that nearest neighbor classifiers predict a specific class label and the probability that this class label is correct for a tuple in non-represented clusters.

Since we are not aware of any information on locations of clusters, but we know that nearest neighbor classifiers return the correct class label for each tuple in represented clusters, we estimate the first probability by the relative number of tuples in represented clusters with class label a_{N1} and a_{N2}, respectively.

The second probability corresponds to the relative number of tuples in the test set which really belong to the predicted class. Since we already consider classification accuracy for tuples in represented clusters in the first component, we reduce the total number of tuples with class label a_{N1} and a_{N2} by the number of tuples with class label a_{N1} and a_{N2} in represented clusters. In order to keep relative frequencies as close to approximate probabilities as possible, we also reduce the total number of tuples in the same way.

This leads to the overall estimation of classification accuracies of nearest neighbor classifiers in focusing contexts which represent well-separated clusterings. □

Proposition 5.1.2 uses the result of proposition 5.1.1 and presents an estimation for classification accuracies of nearest neighbor classifiers in focusing contexts which comprise well-separated clusterings. This estimation is closer to approximate quality of data mining results in terms of classification accuracy since it also contains a component for the amount of correct classifications for tuples in non-represented clusters. The first component of the sum in proposition 5.1.2 is the lower bound for the number of correct classifications from proposition 5.1.1. For all M_R tuples in represented clusters, nearest neighbor classifiers return the true class label. The second component of the sum in proposition 5.1.2 is an estimation of classification accuracy for the remaining $M - M_R$ tuples in non-represented clusters. For each class label, this component combines the probability of classifying a tuple as a member of this class and the probability that this classification is correct.

Note, although this proposition further restricts the focusing context to domains with only two different classes, an extension of proposition 5.1.2 to more than two classes is straightforward.

Probability of Representing Clusters for WSCs

Up to now, we analyzed quality of data mining results of nearest neighbor classifiers without consideration of focusing solutions. Since focusing outputs correspond to training sets, if we apply focusing solutions before data mining, the results of propositions 5.1.1 and 5.1.2 imply strong dependencies between the number of represented clusters in focusing outputs and classification accuracy of nearest neighbor classifiers. Thus, we need to estimate the probability of representing clusters in outputs of focusing solutions, if we want to predict the expected average classification accuracy of nearest neighbor classifiers in combination with focusing solutions. Here, we consider simple random sampling with replacement as the focusing solution.

Proposition 5.1.3 states the probability that simple random samples with replacement of size m represent L_R clusters which contain R_{N1} clusters with class label a_{N1}. In order to justify the definition of $p(L, L_R, R_{N1}, m)$, we use an analogy between the question of representing clusters and traditional example experiments in probability theory. We consider applications of simple random sampling to focusing inputs which represent well-separated clusterings as two-step experiments with distinguishable colored balls. The set of L clusters in C corresponds to L distinguishable balls, and the class label within each cluster determines the color of each ball. Since we deal with two different class labels, the corresponding colors are black and white, for example.

Proposition 5.1.3 (Probability of Representing Clusters for WSCs)

Assume a (database) table (A, T), a class attribute $a_N \in A$ with $dom(a_N) = \{a_{N1}, a_{N2}\}$, a focusing input $f_{in} \subseteq T$ with $M := | f_{in} |$, a focusing output $f_{out} = \text{RANSAM}(f_{in}, m) \subseteq f_{in}$, a well-separated clustering C of f_{in}, a probability function p, $\forall c_l, c_{l'} \in C : p(\text{RANSAM}(f_{in}, 1) = \{t_i\} \wedge t_i \in c_l) = p(\text{RANSAM}(f_{in}, 1)$ $= \{t_{i'}\} \wedge t_{i'} \in c_{l'})$, $C_R \subset C$, $L := \mid C \mid$, $L_R := \mid C_R \mid$, $L_{N1} := \mid \{c_l \in C \mid n_{N1}(c_l) > 0\} \mid$, and $R_{N1} := \mid \{c_l \in C_R \mid n_{N1}(c_l) > 0\} \mid$.

Then, the probability p that f_{out} represents exactly L_R out of L clusters and exactly R_{N1} out of L_R clusters with class label a_{N1} if we draw a simple random sample with replacement of size m is

$$p(L, L_R, R_{N1}, m)$$

$$= \binom{L}{L_R} \cdot \left(\sum_{l=0}^{L_R} (-1)^l \cdot \binom{L_R}{l} \cdot \left(\frac{L_R - l}{L} \right)^m \right) \cdot \frac{\binom{L_{N1}}{R_{N1}} \cdot \binom{L - L_{N1}}{L_R - R_{N1}}}{\binom{L}{L_R}}.$$

Proof:

The probability $p(L, L_R, R_{N1}, m)$ contains two components. The first component estimates the probability of representing L_R clusters, whereas the second component estimates the probability that R_{N1} of L_R clusters belong to class a_{N1}. Both components adopt results from probability theory (Parzen, 1960; Pearl, 1988) in the following way.

The first probability corresponds to the probability that exactly $L - L_R$ of L specified distinguishable balls are not in simple random samples with replacement of size m (see Parzen (1960), page 84).

The second probability matches the probability that simple random samples without replacement of size L_R, which are drawn from populations of L balls that include L_{N1} black balls, contain exactly R_{N1} black balls (see Parzen (1960), page 179). This is the hypergeometric probability law.   □

The first experiment is now an analogy to simple random samples with replacement of size m from an urn which contains L distinguishable balls, L_{N1} black balls and $L - L_{N1}$ white balls. Then, we infer the probability of representing L_R clusters from the probability that simple random samples with replacement of size m do not contain exactly $L - L_R$ of L balls (see Parzen (1960), page 84). The second experiment considers colors of balls in the sample which correspond to class labels of represented clusters. We adopt the *hypergeometric probability law* which represents the probability that simple random samples without replacement of size L_R contain exactly R_{N1} black balls (see Parzen (1960), page 179). Note, since each ball has only a single color, this experiment corresponds to sampling without replacement.

The product of both probabilities is the final probability that simple random samples with replacement of size m represent R_{N1} clusters with class label a_{N1} and $L_R - R_{N1}$ clusters with class label a_{N2}. Note, this analogy between the probability of representing clusters in well-separated clusterings and traditional experiments in probability theory is only possible, if each cluster has the same a-priori probability that simple random samples represent this cluster. For example, this precondition is valid, if each cluster contains the same number of tuples.

Note, it is normally not important to know the number of represented clusters with class label a_{N1} and class label a_{N2}, respectively, but it is sufficient to know the number of overall represented clusters. However, the estimation of classification accuracies (see proposition 5.1.2) uses specific information on the number of represented tuples for each class label. Hence, we need to distinguish between different class labels in represented clusters in order to enable an application of this estimation.

Expected Average Accuracy of Nearest Neighbor Classifiers for WSCs

At this point, we know the approximate classification accuracy of nearest neighbor classifiers for well-separated clusterings in relation to the number of represented clusters in the training set as well as the probability of representing a specific number of clusters, if we apply simple random sampling with replacement as the focusing solution before data mining. In order to estimate the expected average classification accuracy of nearest neighbor classifiers, if we use the output of simple random sampling with replacement as the training set, we now put both results together.

In general, the expected average value of parameters is the sum of the product of specific values and their probability over all possible values (for example, see Parzen (1960), page 343ff). In the context of average case analyses on the relation between focusing contexts and focusing solutions and their focusing success, the parameter is the classification accuracy of nearest neighbor classifiers on well-separated clusterings, and the probability is the probability of representing a specific number of clusters, since the number of represented clusters determines classification accuracy in this context. The set of all possible values corresponds to the set of classification accuracies of nearest neighbor classifiers for all possible numbers R_{N1} of represented clusters with class label a_{N1} and corresponding numbers $L_R - R_{N1}$ of clusters with class label a_{N2}.

Proposition 5.1.4 (Expected Average NN Accuracy for WSCs)

Assume a (database) table (A, T), a class attribute $a_N \in A$ with $dom(a_N) = \{a_{N1}, a_{N2}\}$, a focusing input $f_{in} \subseteq T$ with $M := \mid f_{in} \mid$, a focusing output $f_{out} = \text{RANSAM}(f_{in}, m) \subseteq f_{in}$, a well-separated clustering C of f_{in}, a nearest neighbor classifier NN which uses f_{out} and Sim to classify tuples, $C_R \subset C$, and $T_R := \bigcup_{l=1}^{|C_R|} c_l$. Then, the expected average classification accuracy of NN on f_{in} is

$$\hat{a}(NN, f_{in}, f_{out}, T_R) = \sum_{L_R=0}^{L} \left(\sum_{R_{N1}=0}^{L_R} \tilde{a}(NN, f_{in}, T_R) \cdot p(L, L_R, R_{N1}, m) \right).$$

Proof:

Since the expected value of parameter x is equal to $\sum_x x \cdot p(x)$, possible values for classification accuracy of nearest neighbor classifiers correspond to the result in proposition 5.1.2 for all possible numbers of represented clusters, and the probability that simple random samples with replacement of size m represent L_R clusters, R_{N1} clusters with class label a_{N1}, is $p(L, L_R, R_{N1}, m)$, the expected average accuracy of nearest neighbor classifiers on well-separated clusterings is the definition of $\hat{a}$.   □

In proposition 5.1.4, we define function $\hat{a}$ to compute the expected average classification accuracy of nearest neighbor classifiers on well-separated clusterings. $\hat{a}$ uses the results of proposition 5.1.2 and proposition 5.1.3 to instantiate the general definition of expected values in statistics.

Data mining engineers are now able to use this expected value in three different ways, if the focusing context represents a classification goal, the data characteristics comprise a well-separated clustering, the data mining algorithm is a nearest neighbor classifier, and the focusing solution is simple random sampling with replacement. First, if the data mining engineer specifies the focusing output size in advance, $\hat{a}$ allows an estimation of the expected average quality of data mining results in terms of classification accuracy. On the other hand, the data mining engineer applies $\hat{a}$ for a pre-defined expected classification accuracy in order to compute the required focusing output size.

It is also possible to utilize the result in proposition 5.1.4 in order to estimate the number of well-separated clusters in the focusing context in the following way. We run a number of experiments with simple random sampling with replacement and nearest neighbor classifiers for different focusing output sizes. Then, we compare experimental results with estimations from $\hat{a}$ for different numbers of clusters L. The specific estimation and its concrete value of L, which leads to the best coincidence between experimental results and expected average classification accuracy, yields then an estimation for the number of well-separated clusters in the focusing input.

Ideal Focusing Solutions for WSCs

According to the presented average case analysis, *ideal* focusing solutions for well-separated clusterings and nearest neighbor classifiers represent as many clusters as possible. This means that ideal focusing outputs represent at least as many clusters as they contain tuples, i.e., each tuple up to a number of tuples which corresponds to the number of clusters in the focusing input represents a different cluster. Thus, ideal focusing outputs represent all clusters, if the number of tuples in the focusing output exceeds the number of clusters or is equal to the number of clusters. If ideal focusing outputs contain less tuples than the number of clusters in the focusing input, they still represent the maximum possible number of clusters. Consequently, nearest neighbor classifiers, which use ideal focusing outputs as the training set, predict the correct class label at least for all tuples in a number of represented clusters which corresponds to the number of tuples in the focusing output.

Proposition 5.1.5 (Expected Average NN Accuracy for WSCs II)

Assume a (database) table (A, T), a class attribute $a_N \in A$ with $dom(a_N) = \{a_{N1}, a_{N2}\}$, a focusing input $f_{in} \subseteq T$ with $M := \mid f_{in} \mid$, an ideal focusing output $f_{out} \subseteq f_{in}$ with $m = \mid f_{out} \mid$, a well-separated clustering C of f_{in}, $C_R \subset C$, and $T_R := \bigcup\limits_{l=1}^{|C_R|} c_l$. Then, the expected average classification accuracy of NN on f_{in} is

$$\hat{a}(NN, f_{in}, f_{out}, T_R)$$
$$= \begin{cases} 1, & if\ m \geq L \\ \sum\limits_{R_{N1}=0}^{m} \tilde{a}(NN, f_{in}, T_R) \cdot p(L, m, R_{N1}, m), & else \end{cases}$$

Proof:

The upper case follows from the characteristics of an ideal focusing solution that represents at least as many clusters as it selects tuples in the focusing output and proposition 5.1.1. The lower case is an adaptation of proposition 5.1.4 with a substitution of L_R by m and without the need for the sum over all possible numbers of represented clusters which again follows from the characteristics of an ideal focusing solution. □

Proposition 5.1.5 summarizes these considerations for ideal focusing solutions and applications of nearest neighbor classifiers in focusing contexts which represent well-separated clusterings. If the focusing output contains more tuples than the number of clusters, nearest neighbor classifiers yield maximum classification accuracy of 1, if they use the focusing output of ideal focusing solutions as the training set. If the focusing output includes less tuples, the expected average classification accuracy for nearest neighbor classifiers on well-separated clusterings corresponds to the estimation of the quality of data mining results in terms of classification accuracy for combinations of simple random sampling with replacement and algorithm NN, except that ideal focusing solutions ensure that the focusing output represents as many clusters as the focusing output contains tuples. Hence, we substitute the number L_R of represented clusters by the focusing output size m and do not need to sum over all possible numbers of represented clusters.

Note, although none of the focusing solutions presented in the previous chapter corresponds to an ideal focusing solution, more intelligent sampling techniques are more likely to select an ideal focusing output than statistical sampling techniques, since they take into account similarities and tend to include tuples into the focusing output which belong to different clusters.

5.2   Experimental Validation of Theoretical Claims

In order to validate theoretical claims in the average case analysis for nearest neighbor classifiers on well-separated clusterings in combination with simple random sampling with replacement in comparison to ideal focusing solutions, we perform experiments on artificial domains which represent examples of well-separated clusterings. We compare expected average classification accuracies and experimental results, and infer the usability of the average case analysis in practice from the level of coincidence between both values.

Data Characteristics

The data characteristics of the example domains contain two quantitative attributes and a class attribute with two different labels. We vary the number of well-separated clusters between $L = 4$ and $L = 25$, and each cluster contains the same number of 25 tuples. Maximum similarity between tuples in distinct clusters is always smaller than minimum similarity between tuples within each of these clusters, and each cluster includes only tuples of the same class. Hence, the example focusing contexts represent well-separated clusterings, and each cluster has the same a-priori probability that simple random samples represent this cluster. These characteristics complete the pre-conditions which must be valid in order to apply the average case analysis and its estimations.

Focusing Solution and Data Mining Algorithm

As the focusing solution, we use simple random sampling with replacement and utilize the respective instantiation of generic sampling as its implementation. As the nearest neighbor classifier, we apply inducer IB in the $\mathcal{MLC}++$ library of machine learning algorithms (Kohavi & Sommerfield, 1995; Kohavi et al., 1996). In each focusing context, we run simple random sampling ten times, apply IB to each of the resulting focusing outputs, and use the entire focusing input as the test set. We vary focusing output sizes between $m = 10$ and $m = 100$ in steps of ten tuples. We also estimate the expected average classification accuracy with function $\hat{a}$ (see proposition 5.1.4, page 167) where we plug in values for the number of clusters L and the focusing output size m accordingly.

Experimental Results

Figure 5.2 and figure 5.3 show example results of the average case analysis and its experimental validation in domains with $L = 16$ and $L = 25$ clusters. Both figures depict focusing output sizes in terms of the number of tuples on the x axis and the resulting classification accuracy of IB on the y axis. Dotted lines refer to the expected average classification accuracy, if we apply simple random sampling with replacement before data mining, whereas dashed lines indicate the same estimation for ideal focusing solutions. Each figure also presents experimental

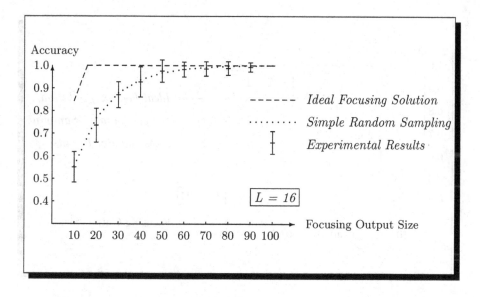

Figure 5.2: Experimental Validation of Average Case Analysis for 16 Clusters

results as error bars which summarize results of each of the ten runs of simple random sampling with replacement and subsequent applications of IB for varying focusing output sizes. Each bar depicts average classification accuracies as well as average values plus and minus standard deviations of ten results as short horizontal lines. We connect horizontal lines which belong to the same focusing output size by vertical lines.

All comparisons between expected average classification accuracies and experimental results show high coincidence between estimations of average classification accuracies and true average classification accuracies with simple random sampling and IB. Hence, we infer that the average case analysis results in appropriate approximations of the quality of data mining results in terms of classification accuracy, if we apply simple random sampling with replacement as the focusing solution before nearest neighbor classifiers as the data mining algorithm in focusing contexts which represent well-separated clusterings.

For example, figure 5.2 indicates that the expected average classification accuracy often matches the true average classification accuracy in experiments with $L = 16$ clusters. In some cases, estimations are slightly too optimistic, and in some cases, they are slightly too pessimistic. The comparison between expected average classification accuracy and experimental results for $L = 25$ leads to lower coincidence but still validates the appropriateness of the average case analysis.

The experimental results also imply lower variances in classification accuracies for increasing focusing output sizes, if we apply simple random sampling

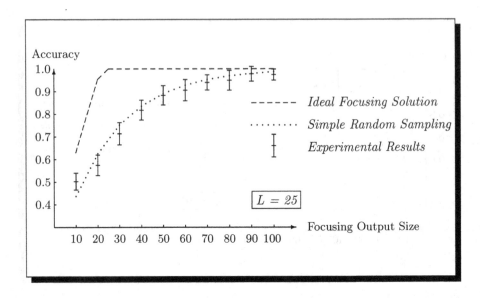

Figure 5.3: Experimental Validation of Average Case Analysis for 25 Clusters

with replacement and IB on the resulting focusing output. This effect meets our expectations since the chance of failing to represent clusters decreases, if the number of tuples in the focusing output increases.

If we compare the expected and true average classification accuracies for simple random sampling with the expected average classification accuracy for ideal focusing solutions, we also see that the behavior of simple random sampling in combination with nearest neighbor classifiers is worse than expected results from applications of ideal focusing solutions. If we apply an ideal focusing solution before nearest neighbor classifiers, we always yield maximum classification accuracy on smaller focusing outputs than in case of simple random sampling. We achieve maximum classification accuracy with ideal focusing solutions exactly as soon as the number of tuples in the focusing output equals the number of clusters in the focusing context. The more clusters the focusing input contains, the more significant is the difference between simple random sampling with replacement and ideal focusing solutions.

In summary, the experiments validate theoretical claims of the average case analysis in focusing contexts which represent well-separated clusterings, if we apply simple random sampling with replacement and nearest neighbor classifiers.

As we mentioned at the beginning of this chapter, average case analyses that cover general relations between focusing contexts and focusing solutions and their focusing success are not feasible. In order to get more insights on relations between focusing contexts and focusing solutions and their focusing success, we present more experimental results in the following chapter.

Chapter 6

Experimental Results

In this chapter, we continue analyses of relations between focusing tasks and focusing solutions and their focusing success. We specify an experimental procedure in order to compare selected focusing solutions in different focusing contexts. We measure focusing success in terms of filter evaluation as well as wrapper evaluation and compare both to each other. At the end, we analyze the experimental results and consolidate most interesting observations as focusing advice. The primary goal of this chapter is further understanding of appropriateness of focusing solutions in relation to focusing contexts.

6.1 Experimental Design

In this section, we outline the experimental design. The experimental design includes an experimental procedure to systematically evaluate focusing solutions in different focusing contexts. We select instantiations of the generic sampling approach to focusing solutions and describe parameter settings for each of the resulting focusing solutions. A set of databases comprises different data characteristics of focusing contexts. All of them define particular classification goals in varying application domains. As data mining algorithms, we use the C4.5 implementation of top down induction of decision trees (Quinlan, 1993) and inducer IB in $\mathcal{MLC}$++ as an implementation of nearest neighbor classifiers (Kohavi & Sommerfield, 1995; Kohavi *et al.*, 1996).

6.1.1 Experimental Procedure

Figure 6.1 shows the experimental procedure. We start each set of experiments with a particular data set. First, we separate the original data set into training set and test set. The training set is a simple random sample from the original data set and contains about 80% of its tuples. The remaining set of tuples

T. Reinartz: Focusing Solutions for Data Mining, LNAI 1623, pp. 173-229, 1999
© Springer-Verlag Berlin Heidelberg 1999

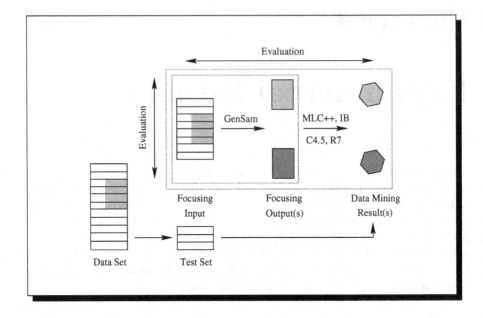

Figure 6.1: Experimental Procedure

constitutes the test set for the classifier application step in classification. The training set is the focusing input for focusing solutions.

Note, although it is common in experimental studies to iterate separations into training set and test set and to use cross-validation in order to get more reliable estimates of classification accuracies, we generate only single separations of the original data set. Since we concentrate experimental studies on evaluation of different focusing solutions and all focusing solutions start with the same focusing input, potential drawbacks of selecting poor training sets by chance have only minor impacts on comparisons of focusing solutions. If the training set is not representative for the classification goal, it is not representative for all focusing solutions in the same way.

6.1.1.1   Focusing Solutions

The next step in the experimental procedure is the application of different focusing solutions to the same focusing input. We apply instantiations of the generic sampling approach to focusing solutions in this focusing step. Specific parameter settings in GENSAM realize different focusing solutions. Since GENSAM allows applications of thousands of different focusing solutions, we are not able to evaluate all possible instantiations. Instead, we select ten different focusing solutions that cover various aspects in generic sampling. We focus experimental studies on statistical sampling techniques and enhancements to the unifying framework of state of the art approaches. In this experimental study, we use the following focusing solutions:

1. *Simple Random Sampling With Replacement*

   (a) without stratification, and (R)

   (b) with stratification (RS)

2. *Systematic Sampling*

   (a) without sorting with random start position, and (S)

   (b) with sorting with fixed start position (1st tuple) (SS)

3. *Leader Sampling*

   (a) without sorting without stratification (L)

   (b) with sorting with stratification (LS)

4. *Similarity-Driven Sampling* with Sorting with Stratification

   (a) with average adaptation strategy without prototype weighting (SA)

   (b) with average adaptation strategy with prototype weighting (SAW)

   (c) with maximum adaptation strategy without prototype weighting (SM)

   (d) with maximum adaptation strategy with prototype weighting (SMW)

Note, all focusing solutions in this experimental study utilize some of our own enhancements to the unifying framework of existing focusing solutions, except simple random sampling without stratification and systematic sampling without sorting with random start position. Both exceptions represent traditional statistical sampling techniques. The rest of focusing solutions either use sorting and stratification to prepare the focusing input before sampling, or they apply more intelligent sampling techniques, leader sampling or similarity-driven sampling.

Leader Sampling

We start the application of focusing solutions with leader sampling. Leader sampling without sorting without stratification and leader sampling with sorting with stratification generate focusing outputs for similarity thresholds $\delta = 0.9$, $\delta = 0.91, \ldots$, and $\delta = 0.99$. Since leader sampling without sorting without stratification is a non-deterministic focusing solution, we repeat applications of pure leader sampling ten times and average evaluation results. In the following, we consider averaged applications of non-deterministic focusing solutions as single experiments. Applications of leader sampling in both variants result in 20 focusing outputs. It depends on the focusing context, if these focusing outputs are really different. In some domains, results of leader sampling are identical, although similarity threshold δ varies.

Similarity-Driven Sampling

We proceed focusing with similarity-driven sampling. We use four different variations of SIMSAM with sorting with stratification. We vary the estimation and adaptation strategy for similarity threshold δ between the average strategy and the maximum strategy. For both strategies, we use similarity-driven sampling with and without prototype weighting.[1] We apply the resulting four focusing solutions without a pre-specified focusing output size first. In these cases, SIMSAM automatically selects focusing outputs of an appropriate size. In addition, we generate focusing outputs with SIMSAM and pre-specified focusing output sizes which correspond to the resulting focusing output sizes from previous runs of leader sampling. In these cases, SIMSAM removes less representative prototypes in order to approximate the pre-specified focusing output size. All in all, similarity-driven sampling generates 84 focusing outputs.

Simple Random Sampling and Systematic Sampling

Once applications of leader sampling and similarity-driven sampling are complete, we continue focusing with two statistical sampling techniques, simple random sampling with replacement and systematic sampling. In order to test the usability of sorting and stratification in combination with focusing solutions in statistics, we also apply simple random sampling with replacement with stratification and systematic sampling with sorting. Moreover, we use random start positions in pure systematic sampling, whereas we fix selection of the first tuple as starting points for systematic sampling with sorting. In this sense, systematic sampling with sorting is a deterministic focusing solution. In all other cases, we again average evaluation results of ten applications of the same focusing solution. We run all four variations of statistical sampling techniques 24 times with pre-specified sample sizes of the previous 24 runs. 20 values result from leader sampling, and 4 values come from the first applications of similarity-driven sampling without pre-specified focusing output sizes. All in all, we get 96 focusing outputs.

Generic Sampling

Table 6.1 summarizes parameter settings in GENSAM for each of the ten focusing solutions. Note, we fix discretization to procedure e_f and maximum stratum size to 100 in case of focusing solutions with stratification. We also settle automatic computation of attribute relevance weights to information gain ratio in case of focusing solutions that depend on attribute relevance.

---

[1]Note, since we did not change data mining algorithms in order to provide usage of prototype weights, we simulate prototype weighting by duplicating each tuple in the focusing output as many times as it represents tuples.

Table 6.1: Parameter Settings in GENSAM

	R	RS	S	SS	L	LS	SA	SAW	SM	SMW
Sorting	-	-	-	✓	-	✓	✓	✓	✓	✓
Stratification	-	✓	-	-	-	✓	✓	✓	✓	✓
Sampling	RanSam	RanSam	SysSam	SysSam	LeaSam	AleSam	SimSam	SimSam	SimSam	SimSam
Discretization	-	e_f	-	-	-	e_f	e_f	e_f	e_f	e_f
max	-	100	-	-	-	100	100	100	100	100
w_j	-	$I_j^+(f_{in})$	-	$I_j^+(f_{in})$	$I_j^+(f_{in})$	$I_j^+(f_{in})$	$I_j^+(f_{in})$	$I_j^+(f_{in})$	$I_j^+(f_{in})$	$I_j^+(f_{in})$
start	-	-	random	1	-	-	-	-	-	-
step	-	-	$\lfloor \frac{f_{in}}{m} \rfloor$	$\lfloor \frac{f_{in}}{m} \rfloor$	-	-	-	-	-	-
δ	-	-	-	-	[0.9;0.99]	[0.9;0.99]	auto	auto	auto	auto
δ *Adaptation*	-	-	-	-	-	-	avg	avg	max	max
w_i	-	-	-	-	-	-	-	✓	-	✓
Deterministic	-	-	-	✓	-	✓	✓	✓	✓	✓

Table 6.2: η Values for Evaluation Criteria

η_σ	η_S	η_V	η_D	η_J	η_T	η_M	η_A	η_C	$\eta^{\leftrightarrow}$	$\eta^{\updownarrow}$
0.5	0.3	0.3	0.3	0.1	0.1	0.2	0.6	0.1	0.25	0.75

6.1.1.2   Data Mining Algorithms

The next step in the experimental procedure is the application of two different data mining algorithms for classification goals. We specialized our attention to top down induction of decision trees and nearest neighbor classifiers in chapter 2. For top induction of decision trees, we use the C4.5 implementation, release 7 (Quinlan, 1993). C4.5 is probably the most widely used implementation of top down induction of decision trees. For nearest neighbor classification, we apply inducer IB in $\mathcal{MLC}++$ (Kohavi & Sommerfield, 1995; Kohavi et al., 1996). The main reasons for this choice are public availability of this tool and identical input formats of training and test set for both C4.5 and $\mathcal{MLC}++$. We use both algorithms with their default settings for parameters.

In the data mining step, we apply both classification algorithms to each of the 200 focusing outputs. Note, for non-deterministic focusing solutions, we run both data mining algorithms on each of the ten focusing outputs and average the quality of the resulting classifiers. Since we chose four non-deterministic and six deterministic focusing solutions, the data mining step includes 938 applications of both C4.5 and IB which results in a total of 1876 classifiers. We estimate classification accuracy of each of these classifiers on the separated test set. Note, for C4.5 we only consider classification accuracies of *pruned* decision trees.

6.1.1.3   Evaluation

The final step in the experimental procedure is the evaluation of the focusing solutions. We use filter evaluation as well as wrapper evaluation, both in isolated and comparative modes. We cumulate values of all filter evaluation criteria and report results of evaluation criterion E_F (see definition 3.4.14, page 77). Note, we set significance level α to 0.1 for all hypothesis tests which corresponds to confidence level 0.9. Similarly, we also cumulate all wrapper evaluation criteria and present values of evaluation criterion E_W (see definition 3.4.18, page 82). Note, we keep results from applications of C4.5 and IB separate. In order to compare filter and wrapper evaluation, we rank the evaluation results for each focusing solution. The best evaluation value gets rank 1, the second best rank 2, and so on. Then, we analyze the relation between ranks in filter evaluation and wrapper evaluation. In order to relate evaluation results for C4.5 and IB, we also compare ranks of wrapper evaluation for both algorithms.

In all evaluation criteria, we punish non-deterministic focusing solutions with $\eta_\sigma = 0.5$. This means that we consider averages between average focusing success and the sum of average focusing success and its standard deviation in ten runs of the same non-deterministic focusing solution. Table 6.2 summarizes η parameter values used in E_F and E_W. In filter evaluation, we equally weigh evaluation criteria that take into account statistical characteristics for each attribute separately, and we add small fractions of evaluation values for joint distributions. In wrapper evaluation, we regard classification accuracy as most important and set its weight to the highest value. We consider storage requirements as slightly more important than efficiency and complexity of data mining results. Finally, we emphasize comparisons among different focusing solutions rather than comparisons between focusing input and focusing output and their respective data mining results. Consequently, the weight for isolated evaluation is smaller than the weight for comparative evaluation.

6.1.2 Data Characteristics

In this section, we summarize most important data characteristics of eight selected databases from the UCI machine learning repository (Murphy & Aha, 1994). The data mining goal classification and these characteristics as well as each of the two data mining algorithms C4.5 and IB constitute focusing contexts in the experimental study. Note, since the experimental procedure is quite complex and needs lots of runs of focusing solutions, data mining algorithms, and evaluation, we use fairly small databases in the experimental study, although this characteristic partly contradicts the original goal of focusing solutions for data mining. However, systematic experimental studies on huge databases are not feasible.

Table 6.3 presents an overview of data characteristics of eight selected databases from the UCI machine learning repository. For each database, table 6.3 shows the number of tuples in the original data set (M), the number of tuples in the training set $(\mid T_{train} \mid)$, the number of tuples in the test set $(\mid T_{test} \mid)$, the redundancy factor of the training set $(R(T_{train}))$, the number of predictive attributes $(N-1)$, the number of qualitative predictive attributes (qual.), the number of quantitative predictive attributes (quant.), the number of classes (N_N), the error rate (see definition 3.4.9, page 68) of applications of C4.5 to the entire training set and applications of resulting classifiers to the test set (C4.5), the error rate of applications of IB to the entire training set and applications of resulting classifiers to the test set (IB), and the default error rate, if we always predict the modus of class labels in the test set (default). In the following descriptions, we refer to each data set by its number (No.) or by its name (Name).

The selected set of databases ensures a wide range of different data characteristics and hence various different focusing contexts for evaluations of different focusing solutions. In particular, these data sets cover the following variations:

Table 6.3: A Selection of UCI Databases

No.	Name	M	$|T_{train}|$	$|T_{test}|$	$R(T_{train})$	$N-1$	qual.	quant.	N_N	C4.5	IB	default
1	Abalone	4177	3342	835	0.0	8	1	7	28	79.4	82.04	80.84
2	Balance	625	500	125	0.0	4	0	4	3	30.4	16.0	44.4
3	Breast	699	560	139	0.01	10	1	9	2	6.5	5.76	30.94
4	Car	1728	1383	345	0.0	6	6	0	4	8.7	7.25	28.12
5	Credit	690	552	138	0.0	15	9	6	2	16.7	21.01	42.03
6	German	1000	800	200	0.0	20	13	7	2	24.0	25.5	29.5
7	Horse	368	295	73	0.0	26	19	7	2	17.8	23.29	38.36
8	Pima	768	615	153	0.0	8	0	8	2	30.7	30.72	37.25

- *Number of Tuples*

  The number of tuples in the original data set varies between 368 (data set 7) and 4177 (data set 1). The number of tuples in the training set deviates between 295 (data set 7) and 3342 (data set 1). These variations allow comparisons of focusing solutions on small, medium, and large focusing inputs.

- *Number of Predictive Attributes*

  The number of predictive attributes varies between 4 (data set 2) and 26 (data set 7). Thus, evaluations of different focusing solutions consider only a few predictive attributes up to many predictive attributes. Consequently, evaluation is able to analyze relations between the number of predictive attributes and focusing success.

- *Attribute Types*

  The selection of UCI databases contains data sets with only qualitative attributes (data set 4), with only quantitative attributes (data sets 2 and 8), and with mixtures of both (data sets 1, 3, 5, 6, and 7). For data sets with mixtures of attribute types, data characteristics cover domains with more qualitative than quantitative attributes (data sets 5, 6, and 7) as well as domains with more quantitative than qualitative attributes (data sets 1 and 3). Consequently, the experimental study allows analyses of relations between attribute types and focusing success.

- *Number of Classes*

  The number of classes varies between 2 (data sets 3, 5, 6, 7, and 8) and 28 (data set 1). Classification tasks with only a few classes are usually easier than classification tasks with many classes. Hence, complexity of selected classification tasks changes from simple tasks to more complex ones. Thereby, we are able to test influences of complexity of classification task on focusing success.

- *Quality of Data Mining Results*

  The selected databases represent classification tasks with good performance (data sets 3 and 4 / data sets 3 and 4), medium performance (data sets 5, 6, and 7 / data sets 2, 5, 6, and 7), and poor performance (data sets 1, 2, and 8 / data sets 1 and 8) of both classification algorithms, C4.5 and IB. Relations between performance of C4.5 and IB also differ among domains. In some examples, C4.5 outperforms IB (data sets 1, 5, 6, 7, and 8), in other examples IB yields lower error rates than C4.5 (data sets 2, 3, and 4). If we take into account default error rates, we observe that the selected data sets contain domains with high potentials of improving default performance by applications of data mining algorithms (data sets 2, 3, 4, 5, and 7) as well as some domains with lower potentials (data sets 1, 6, and 8). All in all, quality of data mining results differs from domain to domain, and we are also able to evaluate relations between data mining success on the entire focusing input and focusing success in terms of data mining success on focusing outputs.

In order to avoid easier focusing tasks with high redundancy, the selected data sets only contain domains with no redundancy, except the *breast* domain with a very small redundancy factor of 0.01.

Note, since we apply two different data mining algorithms on eight different data sets, the experimental study covers 16 different focusing contexts.

6.2 Results and Evaluation

In this section, we present summaries of results in the experimental study. We show evaluation results for focusing success of each of the ten different focusing solutions within each of the 16 varying focusing contexts. First, we discuss filter evaluation and wrapper evaluation for C4.5 and IB separately. Thereafter, we also relate filter evaluation to wrapper evaluation and wrapper evaluation for C4.5 to wrapper evaluation for IB. For all comparisons, we use evaluation criteria E_F and E_W (see definitions 3.4.14, page 77, and 3.4.18, page 82, respectively). We focus our attention on minimum values of evaluation criteria in order to present the best result of each focusing solution. Appendix E also includes results for average and worst focusing outputs of each focusing solution within each focusing context.

6.2.1 Filter Evaluation

Figure 6.2 depicts a bar chart with results for evaluation criterion E_F. The x axis separates the eight different data characteristics, and the y axis indicates minimum values of evaluation criterion E_F. The height of each bar corresponds to the minimum value of evaluation criterion E_F for specific combinations of focusing contexts and focusing solutions. If minimum values of evaluation criterion E_F significantly exceed the range of the majority of values, we depict these significantly higher values with small up arrows and specify true values as explicit numbers. The pattern of each bar corresponds to a specific focusing solution (see legend in figure 6.2).[2]

Table 6.4 generalizes concrete values of evaluation criterion E_F to qualitative ranks of focusing solutions. Whereas bar charts allow analyses of the amount of difference among focusing successes of different focusing solutions, rankings only show qualitative relations.

---

[2]If you prefer bar charts that separate focusing solutions on the x axis and show patterns which correspond to data sets, we refer to appendix E.

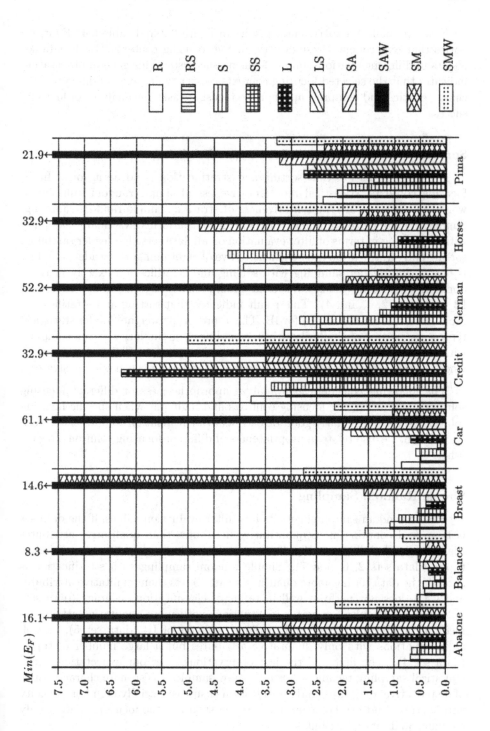

Figure 6.2: Minimum Filter Evaluation

Now, we analyze experimental results in figure 6.2 and table 6.4. First, we notice that experimental results vary among focusing contexts. Each data set results in different relations among focusing successes. This general observation indicates that the selected focusing contexts cover a wide range of different data characteristics and comprise appropriate focusing tasks for analyses of focusing success.

Sorting, Stratification, and Prototype Weighting

Second, we observe that sorting and stratification is advantageous. In all focusing contexts, statistical representativeness of simple random sampling is worse than statistical representativeness of stratified simple random sampling. Similarly, systematic sampling with sorting outperforms systematic sampling without sorting in terms of filter evaluation on all data sets. For leader sampling, sorting and stratification are beneficial in five of eight domains. Leader sampling is better than leader sampling with sorting and stratification exactly in those domains where C4.5 yields worse data mining results in terms of accuracy than IB (data sets 2, 3, and 4). This result indicates that sorting and stratification are more useful for C4.5 than for IB. Third, prototype weights hinder statistical representativeness of similarity-driven sampling. For similarity-driven sampling with average adaptation strategy, this observation is true in all focusing contexts, whereas we perceive two exceptions (data sets 3 and 6) for the maximum strategy.

The following considerations analyze appropriateness of different focusing solutions within varying focusing contexts and highlight when specific focusing solutions show particularly good or bad behavior in relation to data characteristics. Then, we also relate appropriateness of different focusing solutions to each other.

Simple Random Sampling

Simple random sampling gets its best filter evaluation values, if the data set is large and contains more quantitative attributes than qualitative attributes (data set 1). It behaves badly, if the data set includes only a small number of tuples (data sets 2, 6, and 7). Simple random sampling with stratification is good, if the data set has more quantitative attributes than qualitative attributes and C4.5 does not perform well in terms of classification accuracy (data sets 1, 2, and 8). Results are better in situations with more qualitative attributes, medium number of tuples, and medium behavior of IB (data set 5), as well as in situations with only qualitative attributes but a large number of tuples (data set 4). The first and the last observations meet our expectations since statistical sampling techniques usually draw samples to estimate characteristics of quantitative attributes, and stratification ensures especially high homogeneity within strata and low homogeneity between strata, if the focusing context only contains qualitative attributes.

Table 6.4: Rankings of Results with Filter Evaluation

No.	Name	R	RS	S	SS	L	LS	SA	SAW	SM	SMW
1	Abalone	4	3	2	1	9	8	7	10	5	6
2	Balance	8	4	6	1	2	5	7	10	3	9
3	Breast	6	5	4	3	1	2	7	10	9	8
4	Car	5	2	3	1	4	7	8	10	6	9
5	Credit	6	2	3	1	9	8	4	10	5	7
6	German	9	7	8	3	2	1	5	10	6	4
7	Horse	7	5	8	4	2	1	9	10	3	6
8	Pima	5	3	2	1	7	6	8	10	4	9
Average		6.3	3.9	4.5	1.9	4.5	4.8	6.9	10	5.1	7.3

Systematic Sampling

Focusing success of systematic sampling is best, if both data mining algorithms perform badly (data sets 1 and 8). Systematic sampling with sorting is in general a very good focusing solution. It shows its worst result, if the number of attributes is high (data sets 3, 6, and 7). This fact indicates that in case of many attributes sorting is likely to consider only small fractions of all available attributes. This results in lower similarity between tuples that are put close to each other according to relation $\succ$.

Leader Sampling

Leader sampling shows its best results in terms of statistical representativeness, if the data set either contains only a small number of tuples (data sets 2, 3, and 7) or the data set includes many attributes (data sets 6 and 7). Leader sampling with sorting and stratification outperforms all focusing solutions, if the data set has many attributes (data sets 6 and 7). This result indicates that similarity as a key concept in leader sampling contains more reliable information, if it takes into account more attributes. Leader sampling with sorting and stratification also yields good results, if both data mining algorithms perform well and the number of quantitative attributes exceeds the number of qualitative attributes (data set 3).

Similarity-Driven Sampling

The quality of similarity-driven sampling is generally medium or low. Similarity-driven sampling with average adaptation behaves particularly worst, if the data set contains only a small number of tuples (data set 7). Since the average strategy tends to represent many tuples by a single prototype and then only selects small focusing outputs, this observation meets our expectations. Similarity-driven sampling with average adaptation and with prototype weighting is always worst in terms of statistical representativeness. If we use similarity-driven sampling with maximum adaptation, it turns out that two of its best focusing results appear in domains with only quantitative attributes (data sets 2 and 8). This result shows again that information of similarities among tuples is more fine-grained in domains with quantitative attributes than in domains with qualitative attributes. Similarity-driven sampling with maximum adaptation also behaves well on data sets with only a few tuples (data set 7). If we also use prototype weighting along with similarity-driven sampling and maximum adaptation, statistical representativeness again becomes worse.

Simple Random Sampling and Systematic Sampling

Simple random sampling is always worse than systematic sampling in terms of filter evaluation except on data set 7 where simple random sampling is better than systematic sampling, but simple random sampling with stratification is worse than systematic sampling with sorting. Since data set 7 contains the minimum number of tuples, we infer that simple random sampling is more appropriate than systematic sampling in domains with a small number of tuples.

Statistical Sampling and Leader Sampling

If we compare statistical sampling techniques and leader sampling, we detect that simple random sampling is more appropriate than leader sampling, if the performance of both data mining algorithms is bad in terms of accuracy (data sets 1 and 8). We observe the same effect, if potentials for improvement in classification accuracy is large in comparison to default accuracy, but both data mining algorithms only yield medium results (data set 5). We conclude that leader sampling is less appropriate, if the classification task is difficult. In addition, systematic sampling outperforms leader sampling in the same domains as simple random sampling as well as in domains which only contain qualitative attributes (data set 4). In such domains, information of similarity relations among tuples is less fine-grained than in domains with quantitative attributes, since local similarity for qualitative attributes only distinguishes between two values. If we relate enhanced statistical sampling techniques to leader sampling with sorting and stratification, the number of data sets where statistical sampling techniques are worse than leader sampling decreases to three (data sets 3, 6, and 7). Two of these domains (data sets 6 and 7) include many attributes. This result again implies the usefulness of sorting and stratification for statistical sampling techniques.

Statistical Sampling and Similarity-Driven Sampling

Similarity-driven sampling with average adaptation and without prototype weighting is more appropriate in terms of statistical representativeness than simple random sampling in three domains (data sets 2, 5, and 6). In comparison to systematic sampling it is more appropriate in only one domain (data set 6). Similarity-driven sampling with average adaptation and with prototype weighting is significantly worse than all focusing solutions in all focusing contexts. Since we simulate prototype weighting by duplicating prototypes as many times as they represent tuples, similarity-driven sampling with prototype weighting distorts statistical characteristics. Enhanced statistical sampling techniques behave better than similarity-driven sampling with average adaptation in all domains except data set 6 where simple random sampling with stratification is worse than similarity-driven sampling with average adaptation and without prototype weighting.

Similarity-driven sampling with maximum adaptation performs better than similarity-driven sampling with average adaptation in comparison to statistical sampling techniques. If both data mining algorithms yield good results in terms of classification accuracy (data sets 3 and 4), or the number of class labels is large (data set 1), similarity-driven sampling is worse than simple random sampling without and with stratification. Otherwise, similarity-driven sampling outperforms simple random sampling. This result demonstrates that in domains which comprise either simple classification tasks or particularly difficult classification tasks, it is not worth to invest extra efforts of similarity-driven sampling in comparison to simple random sampling. If we use prototype weighting in similarity-driven sampling with maximum adaptation, similarity-driven sampling is only better than simple random sampling in two domains (data sets 6 and 7). However, the difference in domain 7 is less significant.

If we consider systematic sampling and similarity-driven sampling with maximum adaptation, systematic sampling is more appropriate than similarity-driven sampling in the same domains as simple random sampling plus two additional domains (data sets 5 and 8). Similarity-driven sampling with maximum adaptation and with prototype weighting is only better than statistical sampling techniques on data sets 6 and 7. This again shows that prototype weighting is not suitable in terms of statistical representativeness, except if the data set contains many attributes. The relation between enhanced statistical sampling techniques and similarity-driven sampling with maximum adaptation is similar to comparisons between pure statistical sampling techniques and similarity-driven sampling.

Leader Sampling and Similarity-Driven Sampling

If we compare leader sampling with similarity-driven sampling, we see that similarity-driven sampling with average adaptation is only more appropriate than leader sampling in two domains (data sets 1 and 5), and similarity-driven sampling with maximum adaptation is only better than leader sampling in three domains (data sets 1, 5, and 8). Two of these domains (data set 1 and 8) comprise difficult classification tasks where both data mining algorithms achieve bad results in terms of accuracy. The relation between leader sampling with sorting and stratification and similarity-driven sampling with average adaptation is the same as the relation between pure leader sampling and similarity-driven sampling with average adaptation. In contrast, similarity-driven sampling with maximum adaptation outperforms leader sampling with sorting and stratification in two additional domains (data sets 2 and 4). Both domains include only a few attributes. This result implies that similarity-driven sampling is especially more appropriate than enhanced leader sampling in domains with only a small number of attributes.

Average and Maximum Adaptation in Similarity-Driven Sampling

Comparisons between similarity-driven sampling with different adaptation strategies show that the average strategy is only more appropriate in three cases (data sets 3, 5, and 6) in terms of statistical representativeness.

6.2.2   Wrapper Evaluation for C4.5

Figure 6.3 shows a similar bar chart for evaluation criterion E_W for C4.5 as figure 6.2 for evaluation criterion E_F. The x axis again separates eight different focusing contexts, and the y axis indicates minimum values of evaluation criterion E_W for C4.5. The height of each bar corresponds to the minimum value of evaluation criterion E_W for C4.5 for specific combinations of focusing contexts and focusing solutions. The pattern of each bar again corresponds to a specific focusing solution (see legend in figure 6.3).

Table 6.5 generalizes concrete values of evaluation criterion E_W for C4.5 to qualitative ranks of focusing solutions. The analysis of wrapper evaluation follows the same structure as the analysis of filter evaluation.

Again, we first observe that evaluation results differ on different data sets. This general observation implies that the selection of focusing contexts is also appropriate for analyses of focusing success in terms of wrapper evaluation for C4.5.

Sorting, Stratification, and Prototype Weighting

The advantage of sorting and stratification in terms of wrapper evaluation for C4.5 is not as obvious as for filter evaluation. If we compare simple random sampling and stratified simple random sampling, stratification is only advantageous in three domains (data sets 2, 3, and 5). However, differences between wrapper evaluation values for these two focusing solutions are especially small in these domains (see figure 6.3).

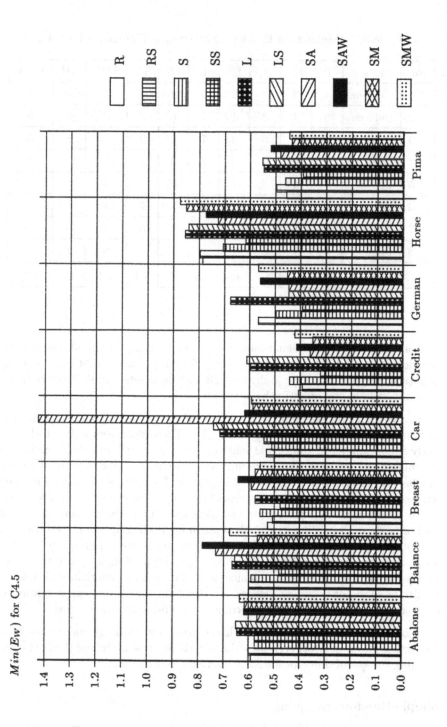

Figure 6.3: Minimum Wrapper Evaluation for C4.5

Table 6.5: Rankings of Results with Wrapper Evaluation for C4.5

No.	Name	R	RS	S	SS	L	LS	SA	SAW	SM	SMW
1	Abalone	4	5	3	2	9	10	1	7	6	8
2	Balance	5	4	3	1	7	6	9	10	2	8
3	Breast	4	3	6	2	7	1	9	10	8	5
4	Car	1	3	2	4	8	9	10	7	5	6
5	Credit	5	4	8	1	9	10	3	6	2	7
6	German	6	9	5	1	10	3	2	7	4	8
7	Horse	5	6	2	1	9	7	3	4	8	10
8	Pima	4	7	5	1	9	10	6	8	2	3
Average		4.3	5.1	4.3	1.6	8.5	7	5.4	7.4	4.6	6.9

Sorting is an useful enhancement to systematic sampling in all domains except for data set 4. Data set 4 is the only focusing context that only contains qualitative attributes. This indicates that sorting is less useful, if the data set does not include any quantitative attributes.

In case of leader sampling, we see advantages of sorting and stratification in four domains (data sets 2, 3, 6, and 7). These data sets either include a fairly small number of tuples and more quantitative attributes than qualitative attributes, or they contain a large number of attributes. We conclude that the number of quantitative attributes is important for success of leader sampling with sorting and stratification in terms of wrapper evaluation for C4.5. Similar to filter evaluation, prototype weighting only increases focusing success of similarity-driven sampling in a single domain for both adaptation strategies. In case of average adaptation, using prototype weights is more appropriate for data set 4, and in case of maximum adaptation, prototype weights prove their usefulness for data set 3. In both domains, both data mining algorithms yield good results in terms of accuracy. Hence, we conclude that prototype weighting is only advantageous in domains that comprise simple classification tasks.

Again, we now analyze specific characteristics of focusing success for each focusing solution in relation to data characteristics separately and then compare focusing success of different focusing solutions to each other.

Simple Random Sampling

Simple random sampling yields its best results on data set 4. This data set contains only qualitative attributes. Since differences between wrapper evaluation values for C4.5 among statistical sampling techniques on this data set are small, we conclude that statistical sampling techniques perform well, if the focusing context does not contain any quantitative attributes. On the other hand, data set 4 contains many tuples, and C4.5 as well as IB achieve good results in terms of accuracy. Hence, a different reasonable explanation for success of statistical sampling in this domain is the size of the data set in terms of the number of tuples and the good performance of both data mining algorithms. In the rest of domains, simple random sampling shows only medium focusing success. Stratified simple random sampling does not necessarily behave better than simple random sampling. It shows its best results also in domain 4 and additionally in domain 3. Since domain 3 is also one of the two domains in the selected focusing contexts where both data mining algorithms perform well, we conclude that simple random sampling is generally appropriate, if the data set comprises simple classification tasks.

Systematic Sampling

Systematic sampling shows its best results in domains 4 and 7. Again, one of these domains contains only qualitative attributes and many tuples, and both data mining algorithms perform well in terms of accuracy. The other domain includes the minimum number of tuples and the maximum number of attributes. Systematic sampling also behaves quite well in domains 1 and 2. If we enhance systematic sampling with sorting, we yield very good focusing success in terms of wrapper evaluation for C4.5, as we already observed for filter evaluation as well. The only exception is domain 4 which contains only qualitative attributes. This result emphasizes our perception that sorting is more useful, if the data set includes quantitative attributes.

Leader Sampling

Focusing success of leader sampling is generally low. In three domains (data sets 2, 3, and 4), it yields slightly more appropriate results. These domains are exactly those domains where IB beats C4.5 in terms of accuracy. Since leader sampling and IB use similarity as their key concept, this observation meets our expectations. Leader sampling with sorting and stratification improves behavior of leader sampling in two of these domains (data sets 2 and 3). It is worse in domain 4 which implies that sorting and stratification is more useful, if the data set contains quantitative attributes. The enhanced leader sampling approach performs best, if both data mining algorithms do well and the data set does not only consist of qualitative attributes (data set 3). Its worst results appear especially in domains where both data mining algorithms achieve bad results (data sets 1 and 8).

Similarity-Driven Sampling

Similarity-driven sampling with average adaptation and without prototype weighting shows its best results in domain 1 which contains many class labels. It behaves also well in domains 5, 6, and 7. In these domains, the number of qualitative attributes exceeds the number of quantitative attributes. Since similarity-driven sampling with average adaptation does not perform well in domain 4, which includes only qualitative attributes, we conclude that similarity-driven sampling benefits from a majority of qualitative attributes but requires also some quantitative attributes. If we also use prototype weighting, quality of similarity-driven sampling with average adaptation decreases except in domain 4. This result implies that prototype weighting is advantageous in domains with only qualitative attributes.

Similarity-driven sampling with maximum adaptation and without prototype weighting achieves its best focusing success in domains 2, 5, and 8. Two of these domains include only quantitative attributes (data sets 2 and 8). Thus, we infer that similarity-driven sampling with maximum adaptation is especially appropriate in domains that only contain quantitative attributes. As for the average adaptation strategy, prototype weighting results in lower focusing success of similarity-driven sampling. Here, the only exception is data set 3. This is the only data set where both data mining algorithms perform well in terms of accuracy but does not only consist of qualitative attributes.

Simple Random Sampling and Systematic Sampling

Simple random sampling yields better wrapper evaluation results for C4.5 than systematic sampling in four domains (data sets 3, 4, 5, and 8). In two of these domains (data sets 3 and 4), both data mining algorithms perform well. This result implies that simple random sampling outperforms systematic sampling in domains with low error rates regardless which data mining algorithm we apply. On the other hand, systematic sampling is especially more appropriate in terms of wrapper evaluation for C4.5, if the data set contains many attributes (data sets 6 and 7). This effect is more significant, if the data set also includes only a small number of tuples (data set 7). If we compare enhanced statistical sampling techniques, we notice that stratified simple random sampling reveals only better evaluation in a single domain (data set 4). This domain contains only qualitative attributes. Hence, we argue that sorting is less advantageous, if the focusing context does not include any quantitative attributes.

Statistical Sampling and Leader Sampling

Both statistical sampling techniques perform better in comparison to leader sampling in all domains. Similarly, enhanced statistical sampling techniques yield higher focusing success than leader sampling with sorting and stratification in six domains. In one domain (data set 3), enhanced leader sampling is better than both enhanced statistical sampling techniques. This is the only domain where both data mining algorithms perform well and which does not only contain qualitative attributes.

Statistical Sampling and Similarity-Driven Sampling

If we compare similarity-driven sampling with average adaptation and without prototype weighting and statistical sampling techniques, we see that similarity-driven sampling is better than statistical sampling in four domains (data sets 1, 5, 6, and 7). Data set 1 includes many different class labels, and data sets 5, 6, and 7 contain more qualitative attributes than quantitative attributes but not exclusively qualitative attributes. The experimental results violate this relation in case of domains with only a small number of tuples (data set 7). In this case, systematic sampling is slightly better than similarity-driven sampling with average adaptation and without prototype weighting. In all other domains, statistical sampling techniques outperform similarity-driven sampling.

If we conduct the same comparison for enhanced statistical sampling techniques, results differ between stratified simple random sampling and systematic sampling with sorting. Whereas similarity-driven sampling with average adaptation and without prototype weighting still outperforms both statistical sampling techniques, if the data set contains many different class labels (data set 1), and similarity-driven sampling is still worse than statistical sampling in domains 2, 3, and 4, we observe differences between random sampling and systematic sampling in domains 5 to 8. Note, IB performs exactly better than C4.5 on data sets 2, 3, and 4. Hence, we conclude that enhanced statistical sampling is more appropriate in terms of wrapper evaluation for C4.5 than similarity-driven sampling with average adaptation and without prototype weighting, if IB outperforms C4.5 in terms of classification accuracy. If C4.5 beats IB on a data set that does not include many class labels (data sets 5, 6, 7, and 8), stratified random sampling is worse than similarity-driven sampling, but systematic sampling with sorting is better than similarity-driven sampling.

If we use similarity-driven sampling with average adaptation and with prototype weighting, statistical sampling is nearly always better than similarity-driven sampling. The only remarkable exception is data set 7 which contains the minimum number of tuples. In this domain, similarity-driven sampling with average adaptation and with prototype weighting outperforms simple random sampling as well as stratified simple random sampling. Hence, we conclude that prototype weighting is advantageous in comparison to simple random sampling, if the data set is small in terms of the number of tuples.

Relations between statistical sampling techniques and similarity-driven sampling with maximum adaptation and without prototype weighting are identical for simple random sampling and systematic sampling. In four domains (data sets 1, 3, 4, and 7), similarity-driven sampling does not yield better wrapper evaluation values for C4.5 than statistical sampling techniques. Whereas the average adaptation strategy performs well in domains with many class labels (data set 1), the maximum strategy fails to beat statistical sampling techniques in such domains. Similarly, similarity-driven sampling with maximum adaptation and without prototype weighting is less appropriate than statistical sampling, if the data set either contains only a few tuples (data set 7), or if both data mining algorithms achieve good results in terms of accuracy (data sets 3 and 4). It is also interesting to note that similarity-driven sampling is especially better than statistical sampling, if the focusing context only contains quantitative attributes (data sets 2 and 8).

In case of domains 1, 3, 4, and 7, comparisons between enhanced statistical sampling and similarity-driven sampling with maximum adaptation and without prototype weighting show the same effects as for pure statistical sampling techniques. In all of these domains, similarity-driven sampling is less appropriate than statistical sampling in terms of wrapper evaluation for C4.5. In the other domains, similarity-driven sampling is still better than stratified random sampling, but systematic sampling with sorting is more appropriate than similarity-driven sampling.

Prototype weighting in similarity-driven sampling with maximum adaptation reveals similar results in comparison to statistical sampling techniques as similarity-driven sampling with average adaptation and with prototype weighting. In most domains, similarity-driven sampling is worse than statistical sampling. Interestingly, the remarkable exception in domain 7 for average adaptation does not hold for maximum adaptation. Instead, maximum adaptation appears particularly useful in domains with only quantitative attributes where both data mining algorithms perform badly (data set 8).

Leader Sampling and Similarity-Driven Sampling

In comparison to leader sampling, similarity-driven sampling without prototype weighting is only generally worse than pure leader sampling and leader sampling with sorting and stratification in a single domain (data set 3). This is the only domain where both data mining algorithms perform well in terms of accuracy and which does not contain only qualitative attributes. In domains 1, 5, 6, 7, and 8, both variants of similarity-driven sampling without prototyping defeat both versions of leader sampling. The only exceptions are data sets 6 and 7 where leader sampling with sorting and stratification is more appropriate than similarity-driven sampling with maximum adaptation. Domains 6 and 7 both contain many attributes. In domains 2 and 4, similarity-driven sampling with average adaptation performs worse than both leader sampling techniques, but similarity-driven sampling with maximum adaptation shows better wrapper evaluation results for C4.5 than leader sampling without and with sorting and stratification. Those domains are exactly domains with only a few attributes.

If we consider leader sampling in comparison to similarity-driven sampling with prototype weighting, we observe that in domains 1, 4, 5, and 8 both similarity-driven sampling approaches beat both leader sampling techniques. In only a single domain (data set 2), both similarity-driven adaptation strategies with prototyping are less appropriate than leader sampling without and with sorting and stratification. This is the only data set with only a small number of exclusively quantitative attributes. In all other domains, we recognize different relations among similarity-driven and leader sampling approaches. In particular, we notice that in domain 7 with the minimum number of tuples similarity-driven sampling with average adaptation defeats maximum adaptation in comparison to leader sampling, if we use prototype weighting.

Average and Maximum Adaptation in Similarity-Driven Sampling

Finally, we compare differences between adaptation strategies in similarity-driven sampling. In three domains (data sets 1, 6, and 7), the average strategy is more appropriate in terms of wrapper evaluation for C4.5 than the maximum adaptation strategy. This observation is true, if we use prototype weighting, or if we do not use prototype weighting. Hence, we conclude that the average strategy is preferable, if the data set either contains many tuples or many attributes. In the rest of domains, the maximum strategy appears better, regardless if we apply prototype weighting or not. The only exception is data set 5 where similarity-driven sampling with maximum adaptation and without prototype weighting is better than its variant with average adaptation, but if we additionally use prototype weights, this relation is opposite.

6.2.3 Wrapper Evaluation for IB

Now, we consider wrapper evaluation for IB. Figure 6.4 shows the corresponding bar chart for evaluation criterion E_W for IB. As in the previous bar charts, the x axis separates eight different focusing contexts, and the y axis indicates minimum values of evaluation criterion E_W for IB. Again, the height of each bar corresponds to minimum values of evaluation criterion E_W for IB for specific combinations of focusing contexts and focusing solutions, and the pattern of each bar corresponds to a specific focusing solution (see legend in figure 6.4). Table 6.6 again generalizes concrete values of evaluation criterion E_W for IB to qualitative ranks of focusing solutions.

As in the experimental study of filter evaluation and wrapper evaluation for C4.5, we observe different relations among focusing success of different focusing solutions. Hence, the selected data sets are also appropriate to analyze wrapper evaluation for IB.

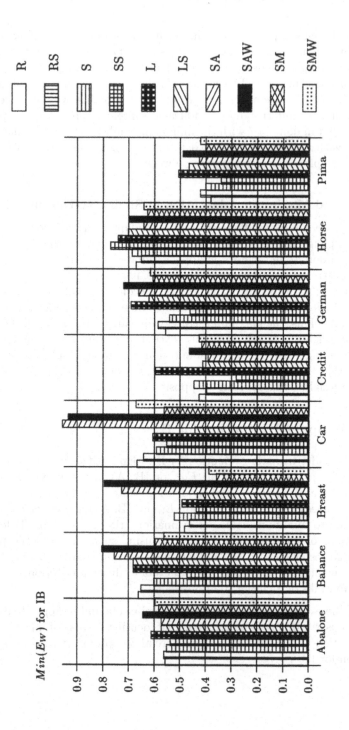

Figure 6.4: Minimum Wrapper Evaluation for IB

Table 6.6: Rankings of Results with Wrapper Evaluation for IB

No.	Name	R	RS	S	SS	L	LS	SA	SAW	SM	SMW
1	Abalone	3	4	2	1	9	6	5	10	7	8
2	Balance	6	5	4	1	8	7	9	10	3	2
3	Breast	6	5	8	4	7	3	9	10	1	2
4	Car	7	6	4	2	5	1	10	9	3	8
5	Credit	7	3	8	1	10	4	2	9	5	6
6	German	3	4	2	1	9	7	8	10	5	6
7	Horse	5	4	6	10	9	8	3	7	1	2
8	Pima	2	6	4	1	10	8	7	9	3	5
Average		4.9	4.6	4.8	2.6	8.4	5.5	6.6	9.3	3.5	4.9

Sorting, Stratification, and Prototype Weighting

In the majority of domains, sorting and stratification is beneficial. Simple random sampling beats stratified simple random sampling only in three domains (data sets 1, 6, and 8). In these domains, IB performs badly in terms of accuracy. This implies that stratification is less useful in domains which comprise difficult classification tasks for nearest neighbor classifiers. Sorting improves systematic sampling in all domains except on data set 7. Since data set 7 contains the minimum number of tuples, we conclude that sorting is less appropriate in terms of wrapper evaluation for IB in domains with only a few tuples. Leader sampling is consistently worse than leader sampling with sorting and stratification in all domains. Since this observation is more significant for wrapper evaluation for IB than for C4.5, we infer that the enhanced leader sampling approach is superior, if we apply nearest neighbor classifiers. Prototype weighting again hinders focusing success of similarity-driven sampling in most domains. The only exceptions, data set 4 in case of average adaptation and data set 2 for maximum adaptation, show only minor variations in the amount of difference between evaluation values (see figure 6.4). These exceptional domains include only a small number of attributes.

Now, we again analyze each focusing solution in relation to focusing contexts separately, and then relate different focusing solutions to each other.

Simple Random Sampling

Simple random sampling yields especially good results in terms of wrapper evaluation for IB, if the quality of data mining results with IB is low (data sets 1, 6, and 8). If we enhance simple random sampling with stratification, we generally observe low focusing success. Hence, we infer that stratification is less appropriate in combination with simple random sampling, if we apply nearest neighbor classifiers.

Systematic Sampling

We see similar effects in experimental results for systematic sampling. This focusing solution is also quite appropriate in situations where IB does not work well on the data set (data set 1, 6, and 8). As we already pointed out in filter evaluation and wrapper evaluation for C4.5, systematic sampling with sorting is generally a very successful focusing solution. A surprisingly significant exception is domain 7 that contains the minimum number of tuples. Hence, we conclude that systematic sampling with sorting is less appropriate in terms of wrapper evaluation for IB in domains with only a few tuples. An additional but less significant exception is domain 3. In this domain, both data mining algorithms perform well in terms of accuracy, but the data set does not only contain qualitative attributes.

Leader Sampling

Although focusing success of leader sampling is generally low, it shows its best performance in domains where IB beats C4.5 in terms of accuracy (data sets 2, 3, and 4). The enhanced leader sampling approach is more appropriate than pure leader sampling. This confirms our expectation that sorting and stratification is useful in combination with leader sampling in terms of wrapper evaluation for IB. This result is especially significant, if the data set either only contains qualitative attributes (data set 4), or both data mining algorithms perform well and the data set does not only include qualitative attributes (data set 3). Leader sampling with sorting and stratification shows its worst results in domains with a large number of attributes (data sets 6 and 7) or in domains with only quantitative attributes (data sets 1 and 8).

Similarity-Driven Sampling

Similarity-driven sampling with average adaptation and without prototype weighting is particularly weak in domains where IB beats C4.5 in terms of accuracy (data sets 2, 3, and 4). If we apply similarity-driven sampling with average adaptation and with prototype weighting, focusing success is low in all domains. Hence, we conclude that prototype weighting is not appropriate in similarity-driven sampling, if we use average adaptation and intend to apply nearest neighbor classifiers.

Similarity-driven sampling with maximum adaptation and without prototype weighting yields its best performance in domains 3 and 7. Data set 3 is the only domain where both data mining algorithms achieve good classification results and which does not contain only qualitative attributes, and data set 4 includes the minimum number of tuples. Moreover, this focusing solution shows quite appropriate results in terms of wrapper evaluation for IB in domains that either have only a small number of attributes (data sets 2 and 4) or include only quantitative attributes (data sets 2 and 8). Prototype weighting does not significantly change focusing success of similarity-driven sampling with maximum adaptation except in domains with only qualitative attributes (data set 4). This result implies that prototype weighting does not hinder similarity-driven sampling, if we use maximum adaptation and nearest neighbor classifiers, and the data set contains quantitative attributes.

Simple Random Sampling and Systematic Sampling

Comparisons between statistical sampling techniques show that simple random sampling outperforms systematic sampling in four domains (data sets 3, 5, 7, and 8), whereas we see the opposite relation in the other domains. If we take into account the amount of difference between evaluation values, most significant differences appear in domains 2, 3, and 4. In these domains, IB defeats C4.5 in terms of accuracy. In domains 2 and 4, simple random sampling is worse than systematic sampling, but in domain 3 it is better. Hence, we argue that systematic sampling is more appropriate in terms of wrapper evaluation for IB, if the data set contains only a few attributes. If we relate enhanced statistical sampling techniques to each other, systematic sampling with sorting clearly behaves better than stratified simple random sampling. The only exception is domain 7 with the minimum number of tuples. This result implies that systematic sampling with sorting is less appropriate in domains with a small number of tuples.

Statistical Sampling and Leader Sampling

Statistical sampling techniques achieve better results than leader sampling in most domains. Exceptions are data sets 3 and 4. In the first exception, simple random sampling is only slightly better than leader sampling, but systematic sampling is worse than leader sampling, and in the other exception, we see only small improvements between systematic sampling and leader sampling. Hence, leader sampling is more suitable in domains where both data mining algorithms perform well in terms of accuracy. Comparisons between enhanced statistical sampling techniques and enhanced leader sampling intensify this insight. Again, statistical sampling techniques are mainly worse in domains where both data mining algorithms work well on the classification task.

Statistical Sampling and Similarity-Driven Sampling

Similarity-driven sampling with average adaptation and without prototype weighting only competes with pure and enhanced statistical sampling techniques in two domains. Similarity-driven sampling is more appropriate than statistical sampling on data sets 5 and 7. The remarkable consequence is good performance in the domain which contains the minimum number of tuples. If we also use prototype weighting in similarity-driven sampling with average adaptation, we observe only one evaluation result where similarity-driven sampling is better than systematic sampling with sorting. Again, this appears in the domain with an exceptionally small number of tuples.

Similarity-driven sampling with maximum adaptation and without prototype weighting performs better than its counterpart with average adaptation in comparison to statistical sampling techniques. It is worse than all statistical sampling approaches in domains 1 and 6, and it is better than all statistical sampling approaches in domains 3 and 7. Again, we notice good performance of similarity-driven sampling in domains with a small number of tuples. Similarity-driven sampling also outperforms most statistical sampling techniques in domains 2 and 4. In these domains, systematic sampling with sorting is the only statistical sampling approach that is better than similarity-driven sampling. However, the difference between evaluation values is very small in domain 4. This means that similarity-driven sampling with maximum adaptation and without prototype weighting is especially favorable in comparison to statistical techniques, if IB yields better classification accuracies than C4.5. Surprisingly, these insights remain valid, if we additionally use prototype weighting. Only in domain 4, the advantage of similarity-driven sampling turns into a disadvantage. This means that prototype weighting hinders similarity-driven sampling with maximum adaptation, if the focusing context only contains qualitative attributes.

Leader Sampling and Similarity-Driven Sampling

Both variants of leader sampling yield more appropriate results than similarity-driven sampling with average adaptation and without prototype weighting in domains 2, 3, and 4. This means leader sampling is better than similarity-driven sampling in domains where IB outperforms C4.5 in terms of accuracy. Leader sampling with sorting and stratification is also able to beat similarity-driven sampling in domain 6. Similarity-driven sampling with average adaptation and with prototype weighting is only superior in comparison to both variants of leader sampling in one domain (data set 7). We infer that prototype weighting does not hinder similarity-driven sampling only in domains with a small number of tuples. If we use maximum adaptation rather than average adaptation in similarity-driven sampling, experimental results imply that similarity-driven sampling is now superior in comparison to pure and enhanced leader sampling in the majority of domains. Prototype weighting again does not distort this effect. The only exception is again the domain which only contains qualitative attributes. In this domain, similarity-driven sampling is worse than both variants of leader sampling.

Average and Maximum Adaptation in Similarity-Driven Sampling

Comparisons of different adaptation strategies in similarity-driven sampling clearly favor maximum adaptation. Similarity-driven sampling with average adaptation is worse than similarity-driven sampling with maximum adaptation in all domains, regardless whether we apply prototype weighting or not. The only exceptions are data sets 1 and 5. In these domains, similarity-driven sampling with average adaptation and without prototype weighting performs better than its counterpart with maximum adaptation. Since differences between both strategies in these domains are very small, we consider these two exceptions as negligible. We conclude that the maximum adaptation strategy in similarity-driven sampling is preferable in terms of wrapper evaluation for IB in comparison to average adaptation.

6.2.4 Comparing Filter and Wrapper Evaluation for C4.5

Whereas we considered filter evaluation and wrapper evaluation for C4.5 and for IB separately in the previous sections, we now turn to comparisons between filter evaluation and wrapper evaluation as well as to comparisons between wrapper evaluation for C4.5 and wrapper evaluation for IB.

Figure 6.5 shows a bar chart for comparisons between evaluation criterion E_F and evaluation criterion E_W for C4.5. This time, the x axis separates ten different focusing solutions. The y axis indicates percentages of focusing outputs which yield better ranks for filter evaluation than for wrapper evaluation for C4.5. Note, whereas we only took into account best focusing outputs for each focusing solution in the previous sections, this bar chart summarizes information of all experiments. For each data set, we rank all focusing outputs according to their filter evaluation value. In the same way, we also rank all focusing outputs according to their wrapper evaluation value for C4.5. Each focusing output has now two different ranks within each focusing context. We compare both ranks for each focusing solution separately, and count the number of focusing outputs which reveal higher ranks for filter evaluation than for wrapper evaluation for C4.5. The height of each bar corresponds to the relative number of better ranks for filter evaluation in percentage for specific combinations of focusing contexts and focusing solutions. The pattern of each bar corresponds to a specific focusing context (see legend in figure 6.5).[3]

---

[3]If you prefer bar charts that separate focusing contexts on the x axis and show patterns which correspond to focusing solutions, we refer to appendix E which contains bar charts for all comparisons between evaluation criteria in this way.

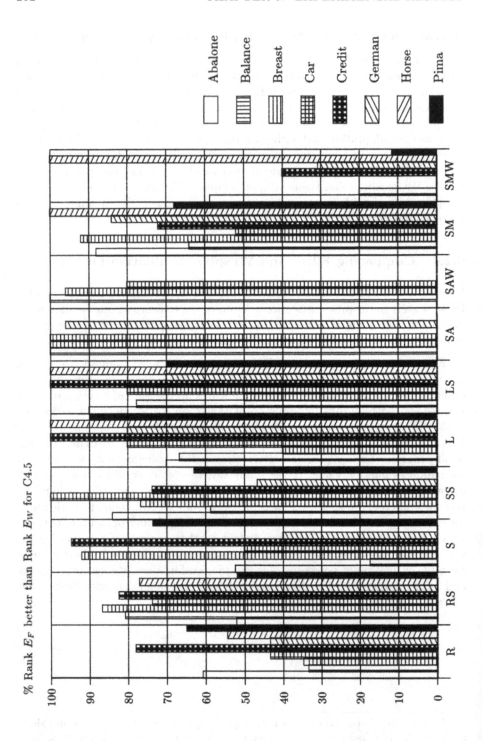

Figure 6.5: Filter Evaluation and Wrapper Evaluation for C4.5

As for separate considerations of experimental results in the previous sections, we structure comparisons between filter and wrapper evaluation along different focusing solutions and comparisons between them. We start comparisons with analyses of influences of sorting, stratification, and prototype weighting, and proceed with each focusing solution separately, before we compare different focusing solutions to each other.

Sorting, Stratification, and Prototype Weighting

Sorting significantly changes relations between filter evaluation and wrapper evaluation for C4.5 in all domains except on data set 7, if we compare systematic sampling and systematic sampling with sorting. Domain 7 is the focusing context with the minimum number of tuples which again indicates that sorting results in less effects in case of a small number of tuples. In four of the rest of domains, sorting largely increases percentages of focusing outputs with higher ranks for filter evaluation than for wrapper evaluation for C4.5. This increase is especially significant in domains 1, 2, and 4. All of these domains include either a large number of tuples or a small number of attributes. Hence, we conclude that sorting as an enhancement to systematic sampling improves statistical representativeness, if the focusing context contains many tuples or a small number of attributes.

Stratification in combination with simple random sampling increases the number of focusing outputs with higher ranks for filter evaluation than for wrapper evaluation in six domains (data sets 2, 3, 4, 5, 6, and 7). This means that stratification improves simple random sampling in terms of filter evaluation in comparison to wrapper evaluation for C4.5. Stratification only results in decreases of higher ranks for filter evaluation in domains 1 and 8. In both domains, both data mining algorithms yield bad performance in terms of classification accuracy. If we apply both preparation steps in generic sampling before leader sampling, the number of higher ranks for filter evaluation than for wrapper evaluation changes in five domains. It increases on data sets 1, 2, and 3, whereas it decreases on data sets 6 and 8.

If we compare similarity-driven sampling without and with prototype weighting in order to analyze effects of prototype weighting on relations between filter evaluation and wrapper evaluation for C4.5, we observe more significant changes in case of maximum adaptation than for average adaptation. Relations between filter and wrapper evaluation only remain the same in domain 7, if we use the maximum adaptation strategy, but additionally keep their values in domains 1, 2, 5, and 8, if we apply average adaptation instead. We infer that prototype weighting has no effect on relations between filter evaluation and wrapper evaluation for C4.5, if the focusing context contains only a small number of tuples, since domain 7 is the data set with the minimum number of tuples. We also conclude that prototype weighting hinders advantages of filter evaluation in comparison to wrapper evaluation, if relations between filter and wrapper evaluation change, since it always results in decreases of the number of focusing outputs with higher ranks for filter evaluation than for wrapper evaluation for C4.5.

Simple Random Sampling

If we inspect relations between filter and wrapper evaluation for simple random sampling, we see four domains with more than half of the focusing outputs with higher ranks for filter evaluation than for wrapper evaluation. In three of these domains (data sets 5, 7, and 8), C4.5 yields better results in terms of classification accuracy than IB. This means that focusing outputs which result from applications of simple random sampling reveal particularly higher ranks for filter evaluation than for wrapper evaluation for C4.5, if C4.5 outperforms IB. Comparisons between filter and wrapper evaluation in case of simple random sampling with stratification show always high percentages of better ranks for filter evaluation than for wrapper evaluation. This implies that stratification pushes focusing success of simple random sampling in terms of filter evaluation. This effect is minimal, if both data mining algorithms perform badly (data sets 1 and 8).

Systematic Sampling

Comparisons between filter evaluation and wrapper evaluation for C4.5 show always better ranks for wrapper evaluation, if we apply systematic sampling and the focusing context contains only a small number of tuples (data set 7). We also observe more higher ranks for wrapper evaluation than for filter evaluation in two additional domains (data sets 2 and 6). There is no preference between filter and wrapper evaluation in domain 4 and almost no preference in domain 1. Since both domains have many tuples, we infer that filter and wrapper evaluation yield similar results for focusing success of systematic sampling in domains with a high number of tuples. In the rest of domains, filter evaluation more often reveals higher ranks than wrapper evaluation for C4.5. If we prepare the focusing input with sorting before systematic sampling, filter evaluation ranks are better than those for wrapper evaluation in the majority of domains. The only minor exception is data set 6, and a significant exception is data set 7.

Leader Sampling

If we apply leader sampling, relations between filter evaluation and wrapper evaluation for C4.5 clearly favor ranks in filter evaluation. For this focusing solution, the only exception is domain 3 with good performance of both data mining algorithms in terms of classification accuracy but not only qualitative attributes. The experimental study indicates similar relations, if we use leader sampling with sorting and with stratification rather than pure leader sampling. In some domains, the number of higher ranks for focusing outputs in terms of filter evaluation increases (data sets 1, 2, and 3), whereas in other domains this number decreases (data sets 6 and 8).

Similarity-Driven Sampling

Similarity-driven sampling with average adaptation without prototype weighting shows most significant results in comparisons between filter evaluation and wrapper evaluation for C4.5. In four domains, focusing outputs always reveal higher ranks in terms of wrapper evaluation (data sets 1, 5, 7, and 8). In two of these domains, both data mining algorithms yield bad performance in terms of classification accuracy. In the remaining four domains, we observe the opposite relation (data sets 2, 3, 4, and 6). In two of these domains, both data mining algorithms perform well in terms of classification accuracy, and in three of these domains IB outperforms C4.5. If we use similarity-driven sampling with average adaptation with prototype weighting, relations between filter and wrapper evaluation change only in three domains. Whereas the difference is negligible in domain 3, it is significant in domain 4, and results in a complete opposite relation in domain 6. Domain 4 is the only focusing context with only qualitative attributes, and domain 6 contains many attributes and many tuples.

If we apply similarity-driven sampling with maximum adaptation, we encounter different and less significant relations between filter evaluation and wrapper evaluation for C4.5. Without prototype weighting, evaluation always yields higher numbers of higher ranks for filter evaluation than for wrapper evaluation. This effect is less significant in domains with only a small number of attributes (data sets 2 and 4). This observation implies that preferences of high ranks for filter evaluation in case of similarity-driven sampling with maximum adaptation without prototype weighting are particularly notable, if the focusing context contains not only a few attributes. If we use prototype weighting, this relation remains only valid in two domains (data sets 1 and 7). These domains include the maximum and the minimum number of tuples. In the rest of domains, similarity-driven sampling shows more focusing success in terms of wrapper evaluation than in terms of filter evaluation.

Simple Random Sampling and Systematic Sampling

If we compare both statistical sampling approaches according to relations between filter evaluation and wrapper evaluation for C4.5, we see that percentages of higher ranks for focusing solutions in terms of filter evaluation are higher for each of these focusing solutions in four domains. Simple random sampling favors filter evaluation more than systematic sampling in domains 1, 2, 6, and 7, whereas systematic sampling yields more higher ranks for filter evaluation than simple random sampling in domains 3, 4, 5, and 8. In two of the latter domains, both data mining algorithms perform well in terms of classification accuracy (data sets 3 and 4), and in one of those domains, the focusing context contains only quantitative attributes (data set 8). For enhanced statistical sampling techniques, relations between simple random sampling and systematic sampling change in three domains (data sets 1, 3, and 5). Now, systematic sampling with sorting results in more higher ranks for filter evaluation than for wrapper evaluation for C4.5 in only three domains (data sets 1, 4, and 8). These domains include either many tuples or only quantitative attributes.

Statistical Sampling and Leader Sampling

Comparisons between statistical sampling techniques and leader sampling show that leader sampling gets more higher ranks in terms of filter evaluation than in terms of wrapper evaluation in all domains except domain 3 where systematic sampling reveals higher percentages of higher ranks. This implies that leader sampling tends to favor statistical representativeness rather than quality of data mining results more often than statistical sampling techniques. If we compare enhanced statistical sampling techniques and leader sampling with sorting and stratification, we mainly observe the same relations. The number of exceptions where statistical sampling reveals higher percentages of higher ranks in terms of filter evaluation than wrapper evaluation for C4.5 now increases to three (data sets 2, 3, and 4). These are exactly those domains where IB beats C4.5 in terms of classification accuracy. Hence, we infer that leader sampling is more appropriate in terms of wrapper evaluation in comparison to filter evaluation than statistical sampling techniques in domains where IB yields good results. This observation meets our expectations since IB and leader sampling both use similarities as a key concept.

Statistical Sampling and Similarity-Driven Sampling

Now, we compare statistical sampling techniques with similarity-driven sampling according to relations between filter evaluation and wrapper evaluation for C4.5. It turns out that similarity-driven sampling with average adaptation never ranks focusing outputs higher in terms of filter evaluation than in terms of wrapper evaluation in domains 1, 5, 7, and 8, regardless whether we apply prototype weighting or not. In two of these domains (data sets 1 and 8), IB behaves better than C4.5 in terms of classification accuracy. In domains 2, 3, and 4, the opposite relation holds between filter and wrapper evaluation, although this effect is less significant, if we use prototype weighting. In conclusion, all statistical sampling techniques favor filter evaluation more than similarity-driven sampling in four domains, whereas the contrary is true in three domains. The only exception is domain 4 where systematic sampling with sorting has higher percentages of higher ranks in terms of filter evaluation than similarity-driven sampling with average adaptation and prototype weighting. This domain contains only qualitative attributes. In one domain (data set 6), applications of prototype weighting reverse relations between statistical sampling techniques and similarity-driven sampling with average adaptation.

If we compare statistical sampling techniques and similarity-driven sampling with maximum adaptation, relations are less significant. The number of focusing outputs with higher ranks for filter evaluation than for wrapper evaluation for C4.5 is higher, if we use similarity-driven sampling with maximum adaptation without prototype weighting than if we apply statistical sampling techniques in domains 1, 2, 4, 6, and 7. These domains contain either many tuples (data sets 1 and 4) or an exceptional number of attributes, either specifically low (data sets 2 and 4) or specifically high (data sets 6 and 7). If we additionally make use of prototype weighting, this relation remains valid only in a single domain (data set 7) with the minimum number of tuples. In the rest of domains, similarity-driven sampling almost always yields higher ranks for filter evaluation than for wrapper evaluation in smaller percentages of focusing outputs than statistical sampling techniques.

For enhanced statistical sampling techniques, comparisons with similarity-driven sampling with maximum adaptation without prototype weighting show higher percentages of higher ranks in terms of filter evaluation than in terms of wrapper evaluation for C4.5 in domains 1, 3, 6, 7, and 8. These focusing contexts include either more quantitative attributes than qualitative attributes (data sets 1 and 3), or many attributes (data sets 6 and 7), or only quantitative attributes and both data mining algorithms perform badly in terms of classification accuracy (data set 8). This indicates the importance of quantitative attributes for focusing success of similarity-driven sampling in terms of statistical representativeness. The same relations between enhanced statistical sampling techniques and similarity-driven sampling according to comparisons between filter and wrapper evaluation are again only valid in domain 7, if we also utilize prototype weighting. This again implies that prototype weighting distorts statistical representativeness.

Leader Sampling and Similarity-Driven Sampling

Relations between leader sampling and similarity-driven sampling with average adaptation are the same, regardless if we apply pure leader sampling or leader sampling with sorting and stratification, as well as regardless whether we use prototype weighting or not. Leader sampling favors filter evaluation in comparison to wrapper evaluation for C4.5 more than similarity-driven sampling in domains 1, 5, 7, and 8. This is the same relation as between statistical sampling techniques and similarity-driven sampling with average adaptation and covers exactly those domains where similarity-driven sampling never yields higher ranks for filter evaluation than for wrapper evaluation. In the rest of domains, similarity-driven sampling with average adaptation without prototype weighting results in a higher number of focusing outputs with higher ranks for filter evaluation than leader sampling. If we use prototype weighting as well, this relation remains valid in domains 2 and 3, and comparisons between filter and wrapper evaluation show identical results for both focusing solutions in domain 4. In all of these domains, IB reveals higher classification accuracy than C4.5.

Similarity-driven sampling with maximum adaptation without prototype weighting yields a higher number of focusing outputs with higher ranks for filter evaluation than both leader sampling approaches only in domains 3 and 6. This relation indicates that leader sampling is more appropriate than similarity-driven sampling with maximum adaptation in many situations. If we utilize prototype weighting in combination with similarity-driven sampling with maximum adaptation, both variants of leader sampling always return more focusing outputs with higher ranks in terms of filter evaluation than in terms of wrapper evaluation, except in domain 7 with the minimum number of tuples. Again, these results indicate that prototype weighting is not appropriate in terms of statistical representativeness.

Average and Maximum Adaptation in Similarity-Driven Sampling

Results of comparisons between both adaptation strategies in similarity-driven sampling are again due to extreme relations between filter and wrapper evaluation for similarity-driven sampling with average adaptation. In domains 1, 5, 7, and 8, none of the focusing outputs from similarity-driven sampling with average adaptation yields higher ranks in terms of filter evaluation than in terms of wrapper evaluation. In domains 2, 3, and 4, the percentage of focusing outputs with higher ranks in terms of filter evaluation for similarity-driven sampling with average adaptation is higher than the respective percentage for similarity-driven sampling with maximum adaptation, regardless whether we apply prototype weighting or not. Domain 6 is the only focusing context where prototype weighting determines the relation between similarity-driven sampling with average and with maximum adaptation.

6.2.5   Comparing Filter and Wrapper Evaluation for IB

In this subsection, we consider relations between filter evaluation and wrapper evaluation for IB. Figure 6.6 shows a bar chart for comparisons between evaluation criterion E_F and evaluation criterion E_W for IB. Again, the x axis separates ten focusing solutions, whereas the y axis and the height of bars indicate percentages of focusing outputs which yield better ranks for filter evaluation than for wrapper evaluation for IB. The pattern of each bar again corresponds to a specific focusing context (see legend in figure 6.5).

Sorting, Stratification, and Prototype Weighting

First of all, we again analyze influences of sorting, stratification, and prototype weighting on relations between filter evaluation and wrapper evaluation for IB. Sorting changes relations between filter and wrapper evaluation in all domains except on data set 6, if we apply systematic sampling. Systematic sampling without sorting favors filter evaluation in comparison to wrapper evaluation for more than half of the focusing outputs in domains 2, 3, 4, 5, and 8. In three of these domains (data sets 2, 3, and 4), IB outperforms C4.5 in terms of classification accuracies. Hence, focusing success of systematic sampling gets higher ranks in terms of filter evaluation than in terms of wrapper evaluation, if IB yields good data mining results. Sorting increases percentages of higher ranks for filter evaluation in four domains (data sets 1, 3, 4, and 7), and decreases them in three domains (data sets 2, 5, and 8). The percentage remains the same in domain 6. Since domains 2 and 8 both include only quantitative attributes, these relations indicate the usefulness of sorting as an enhancement to systematic sampling, if we apply IB and compare filter and wrapper evaluation in domains with quantitative attributes.

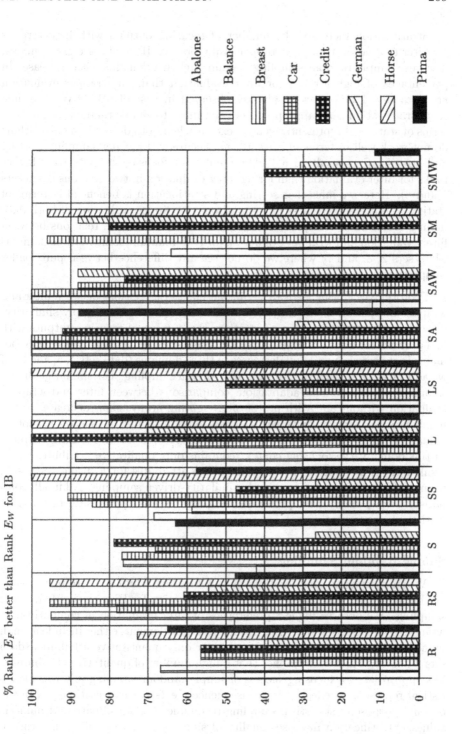

Figure 6.6: Filter Evaluation and Wrapper Evaluation for IB

Stratification increases the number of focusing outputs with higher ranks for filter evaluation than for wrapper evaluation for IB in six of eight domains, if we use simple random sampling. Again, the two domains with decreases in percentages of higher ranks for filter evaluation than for wrapper evaluation are data sets 1 and 8 as we already observed in case of C4.5. We conclude that stratification is an appropriate enhancement to simple random sampling in terms of statistical representativeness, except when both data mining algorithms do not work well in terms of classification accuracy. In leader sampling, sorting and stratification result in a higher number of focusing outputs with higher ranks for filter evaluation than for wrapper evaluation in two domains (data sets 1 and 8). Here, we infer that sorting and stratification is beneficial in terms of statistical representativeness in comparison to wrapper evaluation, if both data mining algorithms yield poor classification results. However, relations between filter evaluation and wrapper evaluation for IB remain the same in three domains (data sets 2, 3, and 7) where we do not see any difference between pure leader sampling and its enhanced variant.

Prototype weighting leads to significant effects on relations between filter and wrapper evaluation in the majority of domains as well. If we apply similarity-driven sampling with average adaptation, the number of focusing outputs with higher ranks for filter evaluation than for wrapper evaluation increases in two domains (data sets 1 and 6) and decreases in three domains (data sets 4, 5, and 8), if we add prototype weighting to similarity-driven sampling. For similarity-driven sampling with maximum adaptation, comparisons between filter and wrapper evaluation without and with prototype weighting always show decreases in the number of focusing outputs with higher ranks for filter evaluation except in domain 4 where evaluation favors wrapper evaluation for all focusing outputs. In this domain, the focusing context contains only qualitative attributes. This result indicates the necessity of quantitative attributes for focusing success in terms of filter evaluation, if we apply similarity-driven sampling. All in all, comparisons again imply that prototype weighting is less appropriate in terms of statistical representativeness.

Simple Random Sampling

We start separate considerations of each focusing solution according to relations between filter evaluation and wrapper evaluation for IB with simple random sampling. We observe highest preferences of filter evaluation in comparison to wrapper evaluation in domains 2, 7, and 8. These domains either include a small number of attributes (data sets 2 and 7) or only quantitative attributes (data sets 2 and 8). This result implies that a high number of qualitative attributes is less appropriate for focusing success of simple random sampling in terms of statistical representativeness. This consequence meets our expectations, since the original purpose of statistical sampling techniques is the estimation of numeric values. Stratification increases quality of simple random sampling in terms of statistical representativeness in many domains, if we consider higher percentages of focusing outputs with higher ranks in terms of filter evaluation than in terms of wrapper evaluation. The only exceptions are domains 1 and 8. In both domains, both data mining algorithms yield bad results in terms of classification accuracy.

Systematic Sampling

Systematic sampling favors filter evaluation rather than wrapper evaluation in five domains (data sets 2, 3, 4, 5, and 8). The smallest percentage of focusing outputs with higher ranks for filter evaluation than for wrapper evaluation appears in domains 6 and 7. Both domains contain many attributes. This observation indicates that systematic sampling is less appropriate in terms of statistical representativeness in comparison to quality of data mining results, if the focusing context includes a large attribute set. This result again coincides with our expectations since statistical sampling techniques are seldomly used to estimate many numeric values at the same time. The enhanced systematic sampling approach yields more than 50% focusing outputs with higher ranks in terms of filter evaluation than in terms of wrapper evaluation in all domains except on data sets 5 and 6. These domains contain more qualitative than quantitative attributes but not only a few tuples. This result again indicates importance of quantitative attributes for focusing success of statistical sampling techniques.

Leader Sampling

Relations between filter evaluation and wrapper evaluation for IB show the minimum number of higher ranks in terms of statistical representativeness in domains 1 and 3, if we apply leader sampling. These domains contain more quantitative attributes than qualitative attributes but not only quantitative attributes. This result means that comparisons between filter and wrapper evaluation favor quality of data mining results, if the focusing context includes a majority of quantitative attributes, and leader sampling is particularly useful in combination with nearest neighbor classifiers in those domains. Relations between filter and wrapper evaluation do not change in three domains (data sets 2, 3, and 7) in case of enhanced leader sampling. Now, leader sampling with sorting and with stratification mainly favors wrapper evaluation in domains 3 and 4. In both domains, both data mining algorithms yield good results in terms of classification accuracy. In the rest of domains, evaluation prefers filter evaluation or there is no preference between filter and wrapper evaluation at all.

Similarity-Driven Sampling

Similar to comparisons between filter evaluation and wrapper evaluation for C4.5, similarity-driven sampling with average adaptation shows most significant results. In three domains (data sets 2, 3, and 4), all focusing outputs from similarity-driven sampling with average adaptation without prototype weighting yield higher ranks for filter evaluation than for wrapper evaluation. In these domains, IB beats C4.5 in terms of classification accuracy. In domains 1 and 7, we observe exactly the contrary. Here, all focusing outputs reveal higher ranks for wrapper evaluation. If we also use prototype weighting in combination with similarity-driven sampling with average adaptation, relations between filter and wrapper evaluation are slightly different. Whereas we see the same relations in domains 2, 3, and 7, and preferences remain similar in domains 1, 4, and 5, we detect reverse relations between filter and wrapper evaluation in domains 6 and 8.

If we inspect relations between filter evaluation and wrapper evaluation for IB for similarity-driven sampling with maximum adaptation without prototype weighting, evaluation favors filter evaluation rather than wrapper evaluation in more than 50% focusing outputs in domains 1, 3, 5, 6, and 7. All of these domains include mixtures of qualitative and quantitative attributes. If we combine similarity-driven sampling with maximum adaptation and prototype weighting, we always count more focusing outputs with higher ranks for wrapper evaluation than for filter evaluation. This relation holds for all focusing outputs in domains where IB is more appropriate than C4.5 in terms of classification accuracy (data sets 2, 3, and 4), or in the domain with the minimum number of tuples (data set 7).

Simple Random Sampling and Systematic Sampling

Comparisons between statistical sampling techniques according to relations between filter evaluation and wrapper evaluation for IB show that simple random sampling yields higher percentages of focusing outputs with higher ranks for filter evaluation in domains 1, 6, 7, and 8. In two of these domains (data sets 1 and 8), both data mining algorithms perform badly in terms of classification accuracy, and in two of these domains (data sets 6 and 7), the focusing context contains many attributes. Comparisons of enhanced statistical sampling techniques change relations between filter and wrapper evaluation in all domains except on data sets 3 and 6. Now, simple random sampling with stratification reveals higher percentages of focusing outputs with higher ranks in terms of filter evaluation than in terms of wrapper evaluation in domains 2, 4, 5, and 6. These data sets include domains with either a small number of attributes or more qualitative attributes than quantitative attributes.

Statistical Sampling and Leader Sampling

The number of focusing outputs with higher ranks for filter evaluation than for wrapper evaluation is higher for leader sampling in comparison to both statistical sampling techniques in five domains (data sets 2, 5, 6, 7, and 8), whereas it is lower in two domains (data sets 1 and 3). This means that leader sampling favors filter evaluation more than both statistical sampling approaches in domains with a small number of attributes and a small number of tuples (data set 2) as well as in domains where C4.5 beats IB in terms of classification accuracy (data sets 5, 6, 7, and 8). The contrary is true in domains with more quantitative attributes than qualitative attributes but not only quantitative attributes (data sets 1 and 3). Comparison between enhanced statistical sampling techniques and leader sampling with sorting and stratification show different relations between filter evaluation and wrapper evaluation for IB. Leader sampling still results in a higher number of focusing outputs with higher ranks for filter evaluation than for wrapper evaluation in domains 6, 7, and 8, but now leader sampling yields a smaller amount of focusing outputs with higher filter evaluation ranks in domains 3 and 4. In both domains, both data mining algorithms perform good in terms of classification accuracy. In domains 1, 2, and 5, relations between enhanced statistical sampling techniques and leader sampling with sorting and stratification according to comparisons between filter and wrapper evaluation depend on the particular statistical sampling approach.

Statistical Sampling and Similarity-Driven Sampling

Statistical sampling techniques without enhancements yield a higher number of focusing outputs with higher ranks in terms of filter evaluation than in terms of wrapper evaluation than similarity-driven sampling with average adaptation without prototype weighting only in domain 1 which includes the maximum number of tuples. This result indicates that statistical sampling techniques are more appropriate in terms of statistical representativeness than similarity-driven sampling, if the focusing context contains many tuples. We observe the opposite relation between statistical sampling and similarity-driven sampling in five domains (data sets 2, 3, 4, 5, and 8). These domains include data sets where IB is better than C4.5 in terms of classification accuracy (data sets 2, 3, and 4). If we utilize prototype weighting in combination with similarity-driven sampling with average adaptation, relations in comparison to statistical sampling techniques remain the same in domains 1, 2, 3, and 4.

Comparisons between enhanced statistical sampling techniques and similarity-driven sampling with average adaptation according to relations between filter evaluation and wrapper evaluation for IB lead to similar conclusions as for pure statistical sampling techniques. A notable exception is domain 7 with the minimum number of tuples. In this domain, simple random sampling significantly increases the number of focusing outputs with higher ranks for filter evaluation than for wrapper evaluation, and systematic sampling completely reverses the relation between focusing success in terms of filter evaluation and in terms of wrapper evaluation. Whereas systematic sampling always reveals higher ranks for filter evaluation, systematic sampling with sorting always yields higher ranks for wrapper evaluation. Consequently, both enhanced statistical sampling techniques now result in higher percentages of focusing outputs, which favor filter evaluation rather than wrapper evaluation, than similarity-driven sampling with average adaptation.

Relations between statistical sampling techniques and similarity-driven sampling with maximum adaptation according to comparisons between filter and wrapper evaluation are different. If we apply statistical sampling techniques without enhancements, we observe a higher number of focusing outputs with higher ranks for filter evaluation in comparison to similarity-driven sampling with maximum adaptation without prototype weighting in three domains (data sets 2, 4, and 8). These domains contain either a small number of attributes or only quantitative attributes. In the remaining five domains, similarity-driven sampling favors filter evaluation more than statistical sampling approaches. If we apply similarity-driven sampling with maximum adaptation with prototype weighting and compare results with statistical sampling techniques, we now see that statistical sampling approaches result in a higher number of focusing outputs with higher ranks for filter evaluation than for wrapper evaluation in almost all situations.

Similar to comparisons between statistical sampling techniques and similarity-driven sampling with average adaptation, relations remain similar, if we now consider enhanced statistical sampling techniques in comparison to similarity-driven sampling with maximum adaptation. Beyond the exception in domain 7 as above, relations between filter evaluation and wrapper evaluation for IB only differ in domain 4. This domain contains only qualitative attributes, and we infer that sorting and stratification are particularly valuable enhancements to simple random sampling and systematic sampling, respectively, if the data set does not contain quantitative attributes.

Leader Sampling and Similarity-Driven Sampling

Again, relations between both variants of leader sampling and similarity-driven sampling with average adaptation are almost the same, regardless whether we apply prototype weighting or not. Leader sampling yields a higher number of focusing outputs with higher ranks for filter evaluation than similarity-driven sampling in domains 1 and 7. In addition, this relation is valid for pure leader sampling in comparison to similarity-driven sampling in domain 5 as well as for leader sampling with sorting and with stratification in comparison to similarity-driven sampling in domain 8. This result implies that similarity-driven sampling with average adaptation favors filter evaluation more than leader sampling only in domains where IB outperforms C4.5 in terms of classification accuracy (data sets 2, 3, and 4), or in domains that contain both many attributes and many tuples (data set 6).

Pure leader sampling as well as enhanced leader sampling result in higher percentages of focusing outputs with higher filter evaluation values than wrapper evaluation values than similarity-driven sampling with maximum adaptation without prototype weighting in four domains (data sets 2, 4, 7, and 8). Two of these domains contain a small number of attributes (data sets 2 and 4), two domains have a small number of tuples (data sets 2 and 7), and two domains include only quantitative attributes (data sets 2 and 8). We infer that leader sampling is superior to similarity-driven sampling with maximum adaptation without prototype weighting in terms of statistical representativeness, if the focusing context comprises either a small number of attributes, a small number of tuples, or only quantitative attributes. If we compare leader sampling and similarity-driven sampling with maximum adaptation with prototype weighting, leader sampling almost always results in more focusing outputs with higher ranks for filter evaluation than for wrapper evaluation. The only exception is data set 1 where this relation does not hold for comparisons between pure leader sampling and similarity-driven sampling.

Average and Maximum Adaptation in Similarity-Driven Sampling

Finally, we consider influences of different adaptation strategies in similarity-driven sampling on relations between filter evaluation and wrapper evaluation for IB. In the first five domains, relations between average and maximum adaptation are the same independent of the application of prototype weighting. Except in domain 1, similarity-driven sampling with average adaptation favors filter evaluation in comparison to wrapper evaluation more often than similarity-driven sampling with maximum adaptation. Since domain 1 contains the maximum number of tuples, we infer that maximum adaptation is more appropriate in terms of statistical representativeness than average adaptation, if the focusing context contains many tuples. In domains 6, 7, and 8, relations between average and maximum adaptation according to comparisons between filter evaluation and wrapper evaluation for IB depend on the application of prototype weighting.

6.2.6 Comparing Wrapper Evaluation for C4.5 and IB

Whereas we considered comparisons between filter and wrapper evaluation for different data mining algorithms in the previous two subsections, we now analyze relations between wrapper evaluation for C4.5 and wrapper evaluation for IB. Figure 6.7 shows a bar chart for comparisons between evaluation criterion E_W for C4.5 and for IB. As in the previous two charts, the x axis separates ten focusing solutions, whereas the y axis and the height of bars now indicate percentages of focusing outputs which yield better ranks for wrapper evaluation for C4.5 than for wrapper evaluation for IB. Again, the pattern of each bar corresponds to a specific focusing context (see legend in figure 6.5).

Sorting, Stratification, and Prototype Weighting

Sorting largely influences relations between wrapper evaluation for C4.5 and wrapper evaluation for IB in six domains. Whereas sorting yields significant increases in the number of focusing outputs with higher ranks for C4.5 than for IB in domains 1, 5, 6, and 8, it results in decreases of preferences for C4.5 in domains 2 and 4. Sorting does not change relations between wrapper evaluations in domains 3 and 7. Decreases appear in domains with only a small number of attributes. This effect indicates that sorting as an enhancement to systematic sampling is less useful for C4.5 in comparison to IB, if the data set does not contain many attributes.

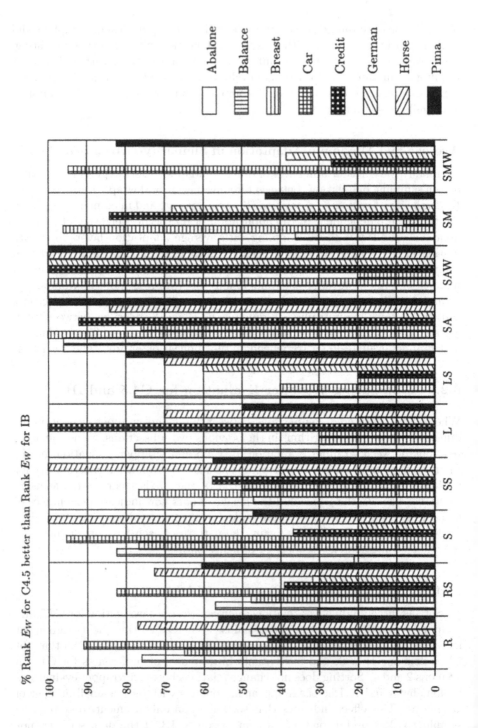

Figure 6.7: Wrapper Evaluation for C4.5 and Wrapper Evaluation for IB

Stratification in combination with simple random sampling always leads to decreases in the number of focusing outputs with higher ranks for C4.5 than for IB except in domain 1 where the relation remains the same and in domain 8 where stratification improves quality of data mining results with C4.5 in comparison to IB. In both domains, both data mining algorithms yield bad results in terms of classification accuracy. Hence, we infer that stratification is not beneficial for C4.5 in comparison to IB unless both data mining algorithms perform badly.

If we apply similarity-driven sampling with average adaptation, prototype weighting decreases the number of focusing outputs with higher ranks for C4.5 than for IB only in domain 4. This domain is the only domain with only qualitative attributes. If we use maximum adaptation rather than average adaptation, prototype weighting is less useful for C4.5 in comparison to IB. It only leads to an increase of the number of focusing outputs with higher ranks for C4.5 than for IB in two domains (data sets 4 and 8).

Simple Random Sampling

Simple random sampling favors C4.5 in less than 50% focusing outputs in domains 1, 5, and 6. Whereas this preference is less significant in domains 5 and 6, we particularly observe this relation in the domain with the maximum number of tuples (data set 1). If we consider simple random sampling with stratification, relations between wrapper evaluation for C4.5 and wrapper evaluation for IB remain similar, but the enhanced approach leads to decreases in the number of focusing outputs with higher ranks for C4.5 than for IB in all domains except on data set 1 and 8. In both domains, both data mining algorithms yield bad results in terms of classification accuracy, and stratification only improves behavior of C4.5 in comparison to IB in one of these domains with only quantitative attributes. This means that stratification is valuable for C4.5 in comparison to IB, if both data mining algorithms perform badly and the focusing context does not contain any qualitative attributes.

Systematic Sampling

Relations between wrapper evaluation for C4.5 and wrapper evaluation for IB show preferences for C4.5 in comparison to IB in more than half of the focusing outputs in four domains (data sets 2, 3, 4, and 7), if we use systematic sampling before data mining. In three of these domains (data sets 2, 3, and 4), IB outperforms C4.5 in terms of classification accuracy, and domain 7 is the data set with the minimum number of tuples. This relation implies that systematic sampling is more appropriate for C4.5 in comparison to IB in domains where C4.5 is less appropriate than IB in terms of classification accuracy, or in domains with only a small number of tuples. Systematic sampling with sorting increases the number of focusing outputs with higher ranks for C4.5 than for IB except in the same domains where pure systematic sampling prefers C4.5 in more than

50% of the cases (data sets 2, 3, 4, and 7). Whereas the number of focusing outputs with higher ranks for C4.5 remains constant in domains 3 and 7, it decreases in domains 2 and 4. These domains contain only a small number of attributes. This means that sorting as an enhancement to systematic sampling is particularly useful for IB in comparison to C4.5 in domains with many attributes.

Leader Sampling

Leader sampling prefers C4.5 more often than IB in three domains (data sets 2, 5, and 7). In two of these domains (data sets 2 and 7), the focusing context contains only a small number of tuples. Hence, leader sampling is particularly beneficial for C4.5 in comparison to IB, if the data set does not contain many tuples. Leader sampling with sorting and with stratification is more appropriate for C4.5 in comparison to IB than pure leader sampling in four domains (data sets 1, 3, 6, and 8), whereas it is less appropriate than pure leader sampling in two domains (data sets 4 and 5). Relations between wrapper evaluation for C4.5 and wrapper evaluation for IB do not change in domains 2 and 7. This result leads to the conclusion that sorting and stratification as enhancements to leader sampling do not make differences according to relations between wrapper evaluation for C4.5 and IB, if the data set contains only a small number of tuples.

Similarity-Driven Sampling

Similarity-driven sampling with average adaptation without prototype weighting fa-vors C4.5 in comparison to IB in all domains except on data set 6. This relation is more significant, if we apply similarity-driven sampling with average adaptation with prototype weighting. In seven of eight domains, all focusing outputs yield higher ranks for C4.5 than for IB. The only exception is data set 4 with only qualitative attributes.

Similarity-driven sampling with maximum adaptation is less appropriate for C4.5 in comparison to IB than similarity-driven sampling with average adaptation. It prefers C4.5 for more than 50% focusing outputs in domains 1, 3, 5, and 6. This observation implies that similarity-driven sampling with maximum adaptation is particularly beneficial for IB in comparison to C4.5 in domains with either only qualitative attributes (data set 4), or with only quantitative attributes (data sets 2 and 8), or with only a small number of tuples (data set 7). If we also apply prototype weighting in similarity-driven sampling with maximum adaptation, we observe a higher number of focusing outputs with higher ranks for C4.5 than for IB only in two domains (data sets 4 and 8).

Simple Random Sampling and Systematic Sampling

If we compare statistical sampling techniques according to relations between wrapper evaluation for C4.5 and wrapper evaluation for IB, we see that simple random sampling prefers C4.5 more often than systematic sampling in four domains (data sets 1, 5, 6, and 8). We infer that systematic sampling is more appropriate for C4.5 in comparison to IB than simple random sampling in domains where IB is better than C4.5 in terms of classification accuracy (data sets 2, 3, and 4) as well as in domains with only a small number of tuples (data set 7). Comparisons between enhanced statistical sampling techniques lead to reverse relations between simple random sampling with stratification and systematic sampling with sorting according to relations between wrapper evaluation for C4.5 and IB in five domains (data sets 1, 2, 4, 5, and 6). Now, simple random sampling favors C4.5 more often than systematic sampling in domains 2, 4, and 8. In two of these domains, the focusing context contains a small number of attributes (data sets 2 and 4). We conclude that sorting as an enhancement to systematic sampling is more appropriate for C4.5 in comparison to IB in domains that have many attributes.

Statistical Sampling and Leader Sampling

Both statistical sampling techniques result in a higher number of focusing outputs with higher ranks for C4.5 than for IB in comparison to leader sampling in three domains (data sets 3, 4, and 7). In two of these domains (data sets 3 and 4), both data mining algorithms yield good results in terms of classification accuracy, and the third domain includes the minimum number of tuples (data set 7). This means that statistical sampling techniques are more appropriate than leader sampling, if we apply C4.5 and both data mining algorithms perform well in terms of classification accuracy, or the data set contains only a small number of tuples. If we inspect relations between enhanced statistical sampling techniques and leader sampling with sorting and with stratification, we observe similar results. Beyond the same data sets as in comparisons between pure statistical sampling techniques and pure leader sampling, both enhanced statistical sampling techniques yield a higher number of focusing outputs with higher ranks for C4.5 than for IB in domain 5 as well.

Statistical Sampling and Similarity-Driven Sampling

Statistical sampling techniques yield a higher number of focusing outputs with higher ranks for C4.5 than similarity-driven sampling with average adaptation without prototype weighting in two domains (data sets 4 and 6), and systematic sampling additionally in domain 7. For systematic sampling, two of these domains have many attributes (data sets 6 and 7). In the rest of domains, similarity-driven sampling results in higher percentages of more focusing outputs with higher ranks for C4.5. This effect is most significant in domains 1, 2, 3, and 8. In two of these domains, the focusing context contains more quantitative attributes than qualitative attributes (data sets 1 and 3), and in two domains, the data set includes only quantitative attributes (data sets 2 and 8). This result implies that similarity-driven sampling with average adaptation without prototype weighting is particularly useful for C4.5 in comparison to IB, if the majority of attributes in the focusing context is quantitative. This effect meets our expectations since similarities distinguish more fine-grained between tuples in case of quantitative attributes. If we also apply prototype weighting, relations between statistical sampling techniques and similarity-driven sampling with average adaptation become more significant. In all domains except on data set 4, similarity-driven sampling always prefers C4.5 rather than IB. Since domain 4 contains only qualitative attributes, this result again strengthens importance of quantitative attributes for focusing success of similarity-driven sampling.

Relations between enhanced statistical sampling techniques and similarity-driven sampling with average adaptation without prototype weighting are the same as relations between pure statistical sampling techniques and similarity-driven sampling except in domains 4 and 7. Systematic sampling with sorting results in a smaller number of focusing outputs with higher ranks for C4.5 than similarity-driven sampling with average adaptation in domain 4, and simple random sampling with stratification shows the same relation in comparison to similarity-driven sampling in domain 7. Since domain 4 contains only qualitative attributes, and domain 7 includes the minimum number of tuples, we infer that sorting as an enhancement to systematic sampling is less useful for C4.5 in comparison to IB, if the data set does not contain any quantitative attribute, and stratification is more powerful in the same respect, if the data set contains only a small number of tuples.

Relations between statistical sampling techniques and similarity-driven sampling with maximum adaptation according to comparisons between wrapper evaluation for C4.5 and wrapper evaluation for IB are different than relations between statistical sampling techniques and similarity-driven sampling with average adaptation. Simple random sampling and systematic sampling yield a higher number of focusing outputs with higher ranks for C4.5 than for IB in comparison to similarity-driven sampling with maximum adaptation without prototype weighting in four domains (data sets 2, 4, 7, and 8). In two of these domains (data sets 2 and 4), the focusing context contains a small number of attributes, in two domains (data sets 2 and 7) a small number of tuples, and in two domains (data sets 2 and 8) only quantitative attributes. If we also apply prototype weighting, both statistical sampling techniques additionally result in higher percentages of preferences for C4.5 than similarity-driven sampling in domains 3 and 5 but no longer in domains 4 and 8.

If we use enhanced statistical sampling techniques and compare their results with similarity-driven sampling with maximum adaptation, we observe the same relations except in domain 1 where systematic sampling with sorting now favors C4.5 more often than similarity-driven sampling. This observation is also true without exceptions for relations between enhanced statistical sampling techniques and similarity-driven sampling with maximum adaptation, if we also utilize prototype weighting.

Leader Sampling and Similarity-Driven Sampling

If we compare focusing success of leader sampling and similarity-driven sampling according to relations between wrapper evaluation for C4.5 and wrapper evaluation for IB, we identify the following results. Pure leader sampling results in a higher number of focusing outputs with higher ranks for C4.5 than similarity-driven sampling with average adaptation without prototype weighting in two domains (data sets 5 and 6). Hence, we conclude that leader sampling without enhancements is only more appropriate for C4.5 in comparison to IB than similarity-driven sampling, if the focusing context contains more qualitative attributes than quantitative attributes but not only a small number of tuples. This result again shows importance of quantitative attributes for similarity-driven sampling approaches. The same relation only appears in domain 4, if we compare leader sampling and similarity-driven sampling with average adaptation with prototype weighting. This observation implies that prototype weighting is less beneficial for C4.5 in comparison to IB in domains with only qualitative attributes. Comparisons between leader sampling with sorting and with stratification and similarity-driven sampling with average adaptation lead to the same relations except in domain 4 where we now observe identical relations between wrapper evaluation for C4.5 and IB for both focusing solutions, as well as in domain 5 where leader sampling no longer exceeds the number of focusing outputs with higher ranks for C4.5.

Relations between both variants of leader sampling and similarity-driven sampling with maximum adaptation without prototype weighting are the same, regardless whether we apply leader sampling with or without enhancements. The only exception is domain 5 where sorting and stratification make a difference. Leader sampling yields higher percentages of focusing outputs with higher ranks for C4.5 in comparison to IB than similarity-driven sampling in four domains (data sets 2, 4, 7, and 8). Again, we point out that these domains represent focusing contexts with only a small number of attributes, or a small number of tuples, or with only quantitative attributes. Pure leader sampling additionally prefers more often C4.5 than similarity-driven sampling in domain 5. If we compare leader sampling and similarity-driven sampling with maximum adaptation with prototype weighting, sorting and stratification change relations between both focusing solutions in domains 5 and 6. In four domains (data sets 1, 2, 3, and 7), both leader sampling approaches favor C4.5 more often than similarity-driven sampling. This relation is also valid between pure leader sampling and similarity-driven sampling in domain 5, and for enhanced leader sampling and similarity-driven sampling in domain 6.

Average and Maximum Adaptation in Similarity-Driven Sampling

Comparisons between alternative adaptation strategies in similarity-driven sampling show higher percentages of focusing outputs with higher ranks for C4.5 than for IB, if we apply average adaptation rather than maximum adaptation without prototype weighting in all domains except on data set 6. Hence, we infer that average adaptation is more appropriate for C4.5 in comparison to IB than maximum adaptation in almost all contexts. This conclusion is also true, if we additionally use prototype weighting. Now, the only exception is domain 4 with only qualitative attributes.

6.3   Focusing Advice

In this section, we consolidate experiences in the experimental study and give some focusing advice that states heuristics for appropriate selections of best suited focusing solutions in relation to focusing contexts. All in all, we notice that focusing success is highly dependent on the focusing context. We do not see general advantages of any focusing solution in comparison to other focusing solutions across all focusing contexts. Instead, specific characteristics in the focusing context determine relations among focusing success of different focusing solutions. Consequently, we structure focusing advice along varying focusing contexts. We specifically separate considerations by concrete implementations of focusing criteria and data mining algorithms (i.e., either E_F, E_W for C4.5, or E_W for IB) as well as individual data characteristics. In the following, we first present general heuristics for utilizations of sorting, stratification, and prototype weighting. Then, we consider focusing solutions in focusing contexts and describe focusing advice for appropriateness of focusing solutions in relation to focusing contexts and specific evaluation criteria.

6.3.1   Sorting, Stratification, and Prototype Weighting

We start focusing advice with the following general rules for applications of sorting, stratification, and prototype weighting:

- E_F

    - Sorting and stratification are generally useful.
    - If C4.5 outperforms IB in terms of accuracy, it is not worth to apply leader sampling with sorting and stratification rather than leader sampling.
    - Prototype weighting generally distorts statistical representativeness.

- E_W for C4.5

    - Sorting is advantageous for systematic sampling, if the data set does not contain only qualitative attributes.
    - Stratification is generally not beneficial for simple random sampling.
    - We recommend leader sampling with sorting and stratification only in domains that either include only a small number of tuples and

more quantitative attributes than qualitative attributes, or the data set contains many attributes.

- Prototype weighting is again generally inappropriate.

- E_W for IB

  - Sorting is a good enhancement to systematic sampling, if the data set does not only contain a few tuples.
  - Stratification is an improvement to simple random sampling unless both data mining algorithms perform badly in terms of classification accuracy.
  - Sorting and stratification are always profitable in leader sampling.
  - Prototype weighting is once more untimely.

6.3.2 Focusing Solutions in Focusing Contexts

In order to state recommendations for specific focusing solutions in relation to individual focusing contexts, we specify particular classes of data characteristics and quality of data mining algorithms. These classes follow up variations within the selected set of UCI databases (see table 6.3, page 180). The left column in the following three tables summarizes these aspects of focusing contexts. We distinguish between the following general classes of focusing contexts:

- *Number of Tuples*

  - M small, i.e., small number of tuples
  - M large, i.e., large number of tuples

- *Number of Predictive Attributes*

  - N small, i.e., small number of attributes
  - N large, i.e., large number of attributes

- *Attribute Types*

  - qual. = 0, i.e., only quantitative attributes
  - quant. = 0, i.e., only qualitative attributes
  - qual. > quant., i.e., more qualitative than quantitative attributes
  - quant. > qual., i.e., more quantitative than qualitative attributes

- *Number of Classes*

  - N_N large, i.e., large number of class labels

- *Quality of Data Mining Results*

  - – C4.5 and IB good, i.e., C4.5 and IB perform well in terms of accuracy

  - – C4.5 and IB bad, i.e., C4.5 and IB perform badly in terms of accuracy

  - – C4.5 bad, i.e., C4.5 performs badly in terms of accuracy

  - – IB bad, i.e., IB performs badly in terms of accuracy

  - – C4.5 better than IB, i.e., C4.5 outperforms IB in terms of accuracy

  - – IB better than C4.5, i.e., IB outperforms C4.5 in terms of accuracy

Focusing Scores

For each evaluation criterion, we review the experimental results and consolidate observations with *scores* for the appropriateness of each focusing solution in relation to data characteristics. We use the following mechanism to compute these scores. First, we consider situations where a specific focusing solution behaves particularly good or particularly bad for each focusing solution separately. We add a *plus* for good focusing success and assign a *minus* for poor performance of focusing solutions. Thereafter, we analyze comparisons between different focusing solutions. For each pair of comparisons, we relate focusing success of the first focusing solution to focusing success of the second. The superior solution again gets a plus, whereas the inferior solution deserves a minus. At the end, we sum up all plus and minus scores which results in final scores for each focusing solution.

Tables 6.7, 6.8, and 6.9 summarize scores for each focusing solution in relation to data characteristics for filter evaluation, wrapper evaluation for C4.5, and wrapper evaluation for IB, respectively. Note, in individual cases, scores abstract from specific experimental results but keep most important relations among focusing solutions.

If we now plan to apply focusing solutions in KDD projects, we use these tables and decide which focusing solution is preferable dependent on the data characteristics in the present focusing context. All of these tables comprise heuristics for appropriateness of focusing solutions and enable careful selections of best suited focusing solutions. Rows in each table allow comparisons among focusing solutions for specific data characteristics, whereas columns relate concrete focusing solutions across varying data characteristics. Negative scores mean that this focusing solution is generally less appropriate in this context, whereas positive scores recommend applications of this focusing solution. The lower the score is, the worse is the focusing solution, and the higher the value is, the better is the focusing solution. Lowest scores correspond to worst focusing solutions, and highest values refer to best focusing solutions.

Table 6.7: Focusing Advice for Filter Evaluation

Context	R	RS	S	SS	L	LS	SA	SAW	SM	SMW
M small	0	+1	-1	+4	+5	0	-4	-5	+1	-4
M large	-1	+1	+2	+5	+4	0	-3	-5	+1	-4
N small	-1	+1	+2	+5	+4	-2	-4	-5	+3	-4
N large	-3	-3	-1	0	+5	+6	-4	-5	+1	+4
qual. $= 0$	-1	+1	+1	+5	+4	0	-3	-5	+1	-5
quant. $= 0$	-1	+2	+3	+5	+2	0	-3	-5	+1	-4
qual. $>$ quant.	-1	+1	+1	+5	+4	0	-3	-5	+1	-4
quant. $>$ qual.	-1	+1	+1	+5	+4	0	-4	-5	+1	-4
N_N large	+3	+3	+3	+5	-1	-4	+1	-5	-1	-4
C4.5 and IB good	+1	+3	+3	+5	+4	0	-3	-5	-5	-4
C4.5 and IB bad	+1	+2	+3	+5	-2	-2	-3	-5	+5	-4
C4.5 bad	-1	+1	+1	+5	+4	0	-3	-5	+1	-4
IB bad	-1	+1	+1	+5	+4	0	-3	-5	+1	-4
C4.5 better than IB	-1	+1	+1	+5	+4	0	-3	-5	+1	-4
IB better than C4.5	-1	+1	+1	+5	+4	0	-3	-5	+1	-4
Average	-0.6	+1.1	+1.4	+4.6	+3.4	-0.1	-3.0	-5.0	+0.9	-3.5

Table 6.8: Focusing Advice for Wrapper Evaluation for C4.5

Context	R	RS	S	SS	L	LS	SA	SAW	SM	SMW
M small	0	0	+4	+3	-2	-2	-3	-1	-3	+5
M large	+3	+3	+3	+5	-2	-2	-1	-3	+1	-5
N small	+3	+3	+3	+5	-1	-1	-5	-5	+3	-3
N large	+2	+3	+5	+5	-2	-1	-1	-3	0	-5
qual. = 0	+1	0	+3	+4	-2	-2	-3	-5	+4	+5
quant. = 0	+4	+5	+3	+2	-2	-3	-4	-4	+3	-3
qual. > quant.	+1	+3	+2	+5	-2	-2	+2	-5	+3	-3
quant. > qual.	+3	+3	+3	+5	-2	-2	-3	-5	+3	-3
N_N large	+1	+2	+2	+4	-2	-2	+4	-5	+3	-3
C4.5 and IB good	+6	+3	+3	+3	0	+5	-5	-5	-5	-2
C4.5 and IB bad	+1	0	+2	+3	-2	-3	-3	-5	+3	+5
C4.5 bad	+3	+3	+3	+5	-2	-2	-3	-5	+3	-3
IB bad	+3	+3	+3	+5	-2	-2	-3	-5	+3	-3
C4.5 better than IB	+3	+2	+3	+5	-2	-2	-2	-5	+3	-3
IB better than C4.5	+3	+4	+3	+6	-1	-1	-5	-5	+3	-3
Average	+2.4	+2.5	+3.1	+4.3	-1.7	-1.5	-2.3	-4.4	+1.9	-1.6

Table 6.9: Focusing Advice for Wrapper Evaluation for IB

Context	R	RS	S	SS	L	LS	SA	SAW	SM	SMW
M small	-2	0	-1	-5	-4	-5	+4	-1	+8	+7
M large	+3	+2	+3	+5	-3	-2	-5	-6	+2	+2
N small	0	0	+2	+4	-3	-2	-5	-6	+6	+6
N large	+3	+2	+3	+5	-3	-3	-5	-6	+2	+2
qual. = 0	+3	+2	+3	+5	-3	-3	-5	-6	+3	+3
quant. = 0	+2	+1	+4	+6	+2	+4	-5	-6	-2	-6
qual. > quant.	+3	+2	+3	+5	-3	-2	-5	-6	+2	+2
quant. > qual.	+3	+2	+3	+5	-3	-2	-5	-6	+2	+2
N_N large	+3	+2	+3	+5	-3	+1	-5	-6	0	0
C4.5 and IB good	+2	0	0	+3	+1	+3	-5	-6	+3	+2
C4.5 and IB bad	+3	+2	+3	+5	-3	-3	-4	-6	+2	+2
C4.5 bad	+3	+2	+3	+5	-3	-2	-5	-6	+2	+2
IB bad	+4	+3	+3	+5	-3	-2	-5	-6	+2	+2
C4.5 better than IB	+2	+3	+3	+5	-3	-2	-5	-6	+2	+2
IB better than C4.5	+4	+3	+2	+5	-1	-1	-8	-6	+2	+2
Average	+2.4	+1.7	+2.5	+4.2	-2.3	-1.4	-4.5	-5.7	+2.4	+2.0

Focusing Advice for Filter Evaluation

Table 6.7 depicts scores of focusing solutions in relation to focusing contexts for filter evaluation. Average values (at the bottom) indicate a central tendency that statistical sampling techniques yield sufficient focusing success in terms of statistical representativeness. Since the original purpose of statistical sampling techniques is the estimation of characteristic values, this tendency confirms appropriateness of statistical sampling techniques in terms of filter evaluation. Random sampling is slightly worse, whereas systematic sampling with sorting is clearly better. We also notice that leader sampling does surprisingly well, whereas its enhanced variant is not recommended. Similarity-driven sampling is generally only appropriate in individual cases. Average adaptation is clearly worse than maximum adaptation. Since average adaptation is not as strict in selection of prototypes as maximum adaptation, and average adaptation tends to select fewer prototypes with less variations in characteristic values than maximum adaptation, this observation meets our expectations.

Focusing Advice for Wrapper Evaluation for C4.5

Table 6.8 contains scores of focusing solutions in relation to focusing contexts for wrapper evaluation for C4.5. First, we notice that these scores are different from scores for filter evaluation. This difference strengthens the general impression that focusing success depends on the specific evaluation criterion. On average, statistical sampling techniques outperform more intelligent sampling techniques in terms of wrapper evaluation for C4.5. This average result implies that estimations of proportions which lead to estimations of attribute value frequencies are superior, if we apply statistical sampling techniques. Since attribute value frequencies mainly guide top down induction of decision trees, this result confirms our expectations. Similar to filter evaluation, we recognize significant success of systematic sampling with sorting, and preferences for maximum adaptation rather than average adaptation. We also recognize that simple random sampling is more appropriate in domains with many tuples rather than only a few tuples, or on data sets with a small number of attributes rather than many attributes. In these domains, random sampling is less likely to miss important information. Thus, this relation between random sampling and the number of tuples and attributes verifies our hypotheses.

Focusing Advice for Wrapper Evaluation for IB

Finally, table 6.9 depicts scores of focusing solutions in relation to focusing contexts for wrapper evaluation for IB. Again, statistical sampling techniques yield surprisingly high focusing success. As for wrapper evaluation for C4.5, we also notice that both statistical sampling techniques do better in case of many tuples but now favor a large number of attributes. In addition, simple random sampling is more appropriate in domains with only quantitative attributes. If we apply nearest neighbor classifiers, similarity-driven sampling with maximum adaptation significantly gains focusing success. Since the usage of similarities is common to similarity-driven sampling and nearest neighbor classifiers, and maximum adaptation retains more information than average adaptation, this circumstance coincides with our expectations. On the other hand, if we compare results of leader sampling and similarity-driven sampling with maximum adaptation, it is advantageous to retrieve the most similar prototype and to dynamically adapt similarity thresholds.

Summary

In summary, we are astonished about focusing success of systematic sampling with sorting. We infer that close relations between sorting and stratification as well as impacts of sorting on similarities make systematic sampling with sorting a very good focusing solution in general. In almost all situations across all evaluation criteria, systematic sampling with sorting is a good starting point for focusing in KDD. An additional advantage of this focusing solution as well as all deterministic focusing solutions is the exclusion of randomness in selections of focusing outputs. Non-deterministic focusing solutions always incorporate dangers of generating worst focusing outputs.

Comparisons of focusing success in different domains as well as comparisons between results with different evaluation criteria strengthen the important influence of the focusing context. Appropriateness of focusing solutions depends on data mining goal, data characteristics, and data mining algorithm, as well as the specific evaluation criterion that realizes the focusing criterion. An appropriate focusing solution in terms of statistical representativeness is not necessarily also appropriate according to quality of data mining results, and vice versa a single focusing solution often yields good results in terms of wrapper evaluation but fails to have focusing success with respect to filter evaluation. Likewise, we also observe differences in appropriateness of focusing solutions, if we compare different domains or different data mining algorithms.

At this point, we also emphasize that both evaluation criteria, E_F and E_W, simultaneously consider all aspects of filter evaluation and wrapper evaluation, respectively. If we only take into account single aspects, results and focusing success vary as well. For example, if we regard classification accuracy as the only relevant component in wrapper evaluation, we often observed higher classification accuracies, if we apply more intelligent sampling techniques and use resulting focusing outputs as training sets for classification, than if we use the entire focusing input as the training set.

Chapter 7

Conclusions

In this chapter, we summarize contents of this dissertation and recapitulate its main contributions. We also outline more related work for different focusing tasks, describe limitations of our work, and raise several issues for future work. Some closing remarks finish this dissertation.

7.1 Summary and Contributions

In summary, this dissertation developed, analyzed, and evaluated focusing solutions for data mining. After an introduction to KDD and data mining as well as an outline of general goals in focusing, we started with more detailed descriptions of KDD processes. We introduced four classes of humans that are usually involved in KDD projects: Business analysts, data mining engineers, application developers, and end users. Then, we outlined a process model for KDD that contains seven different phases: Business understanding, data understanding, data preparation, data exploration, data mining, evaluation, and deployment. Focusing is a particular task in data preparation, and we characterized relations between data selection in data preparation and focusing.

Thereafter, we discussed three aspects in KDD that mainly influence the definition of focusing tasks and the development or selection of appropriate focusing solutions: Data mining goals, data characteristics, and data mining algorithms. We considered phases before and after data preparation in KDD, and we described each of these components in focusing contexts in more detail. At the end of the second chapter, we selected classification with qualitative class attributes as the primary data mining goal and chose two specific data mining algorithms, top down induction of decision trees and nearest neighbor classifiers, for further examination.

Since focusing in general was too generic, the third chapter refined the overall set of focusing tasks and selected a particular subset of focusing tasks for further

T. Reinartz: Focusing Solutions for Data Mining, LNAI 1623, pp. 231-238, 1999
© Springer-Verlag Berlin Heidelberg 1999

consideration. We presented a framework for the definition of focusing tasks that consists of focusing specifications and focusing contexts. A focusing specification includes definitions of focusing input, focusing output, and focusing criterion. A focusing context contains data mining goal, data characteristics, and data mining algorithm. We discussed each of these six concepts and their relations to each other in detail.

The definition of specific filter and wrapper evaluation criteria for focusing success in isolated and comparative modes as well as cumulations of these criteria to take into account more than single evaluation aspects completed the selection of specific focusing tasks. At the end of the first block, we chose focusing tuples to simple subsets or constrained subsets according to representativeness in the sense of filter and wrapper evaluation as the specific subset of focusing tasks for further examination. This focus completed the goal definition for this dissertation.

At the beginning of the second block, we started the development, analysis, and evaluation of focusing solutions for data mining with analyses of existing state of the art efforts towards focusing solutions. We identified a unifying framework that consists of three steps: Sampling, clustering, and prototyping. This framework covers all existing work as instantiations. We described each step in more detail and discussed several example instantiations: Windowing, dynamic sampling using the PCE criterion, leader clustering, CLARANS, instance-based learning, and prototype construction in PL. For each approach, we characterized its main advantages and disadvantages and related each example instantiation to the unifying framework.

The development of more intelligent sampling techniques followed up the unifying framework of focusing solutions and introduced several enhancements that reuse positive aspects of existing work, whereas they overcome their negative properties. The general approach in more intelligent sampling follows up leader clustering and assumes the existence of subsets of sufficiently similar tuples which are represented by a small set of prototypes. The most important enhancements include: Sorting and stratification as two additional preparation steps, automatic estimation and adaptation of similarity thresholds in leader sampling, and attribute relevance measures.

We put all enhancements together in a unified approach to focusing solutions. We implemented the unified approach as a generic sampling algorithm and integrated this algorithm into a commercial data mining system, CLEMENTINE. This integration facilitates easy usage of different focusing solutions and straightforward integration of focusing solutions into KDD processes. At the end of the fourth chapter, we achieved the general goal of this dissertation, contributing to the focusing component of a knowledge discovery system.

Analysis and evaluation of focusing success of several focusing solutions in varying focusing contexts started with an average case analysis which estimates expected average classification accuracies in specific domains in case of nearest neighbor classifiers in combination with simple random sampling with replacement and in comparison to ideal focusing solutions. First experiments validated theoretical claims in this analysis.

In order to provide more comparisons for additional focusing solutions and additional focusing contexts, we set up an experimental procedure and selected ten specific instantiations of generic sampling as well as eight different data sets from the UCI repository. The experimental procedure enabled systematic comparisons of these focusing solutions in each of these focusing contexts. We chose simple random sampling with replacement without and with stratification, systematic sampling without and with sorting, leader sampling without and with stratification and sorting, and similarity-driven sampling without and with prototype weighting and different estimation and adaptation strategies.

Experimental results confirmed strong dependencies between focusing contexts and focusing success of focusing solutions. We consolidated experiences in the experimental study and described heuristics for appropriate selections of best suited focusing solutions in relation to specific focusing contexts as focusing advice. In summary, we were astonished about focusing success of systematic sampling with sorting. In almost all situations across all evaluation criteria, systematic sampling with sorting is a good starting point for focusing in KDD.

Main Contributions

The main contributions of this dissertation include the following issues:

- *KDD Process*

  In the context of this dissertation, we improved the understanding of the overall KDD process and its deployment into industrial businesses. Although this issue was not the main focus of this dissertation and we only outlined the resulting process model, we currently follow up these concerns in a separated project. A consortium of leading data mining suppliers and large industrial organizations started the project CRISP-DM (CRoss-Industry Standard Process for Data Mining). The overall goal in CRISP-DM is the development of an industry-neutral and tool-independent standard process model for data mining which guides concrete data mining projects across all businesses.

- *Focusing Tasks*

  One of the first major contributions in this dissertation was the careful analysis of potential focusing tasks and the definition of a framework for identification and specification of concrete focusing tasks in KDD. In the future, this framework helps the research community to precisely state which focusing task they solve and supports practitioners to specifically identify their focusing tasks in the context of their KDD projects. This makes comparisons of existing approaches in different fields more feasible. This framework also supports systematic analyses of advantages and disadvantages of existing efforts and combinations of state of the art solutions.

- *Unifying Framework of Focusing Solutions*

  We identified a unifying framework of existing state of the art efforts towards focusing solutions. This framework covers each existing approach as an instantiation. Again, this allows researchers and practitioners to compare approaches in a unique way. This framework also helps to identify potential weaknesses and existing strengths of state of the art approaches towards focusing solutions.

- *More Intelligent Sampling Techniques*

  Enhancements to the unifying framework lead to more intelligent sampling techniques. These novel approaches proved to outperform less intelligent focusing solutions in some domains. All in all, the enhancements show new directions for potentials of focusing in KDD, and some of them might also be integrated into other existing focusing solutions. For example, many approaches benefit from utilizations of sorting and stratification as additional preparation steps.

- *Generic Sampling*

  The unification of all enhancements in generic sampling provides numerous different focusing solutions. Each specific parameter setting corresponds to a single concrete focusing solution. The implementation of generic sampling in a commercial data mining system ensures easy applications of focusing solutions for all types of users and straightforward integrations of focusing solutions into KDD processes.

- *Average Case Analysis*

  The average case analysis for specific classification goals and nearest neighbor classifiers in combination with simple random sampling as well as its experimental validation provide first insights in relations between focusing tasks and focusing solutions and their focusing success.

- *Experimental Results*

  The systematic experimental study and its results lead to additional heuristics and focusing advice that guide data mining engineers in selecting most appropriate focusing solutions in relation to specific focusing contexts. Thus, focusing advice provides further insights in relations between focusing tasks and focusing solutions and their focusing success.

Note, in addition to the work on focusing tuples, attribute relevance measures and discretization techniques also contribute to focusing attributes and focusing values, respectively.

In comparison to the initial goals of this dissertation which we stated in the introductory chapter, we achieved each of these goals. The process model for KDD, which we outlined in the second chapter, helped to understand the KDD process and to identify relevant aspects that mainly influence the definition of focusing tasks and the development or selection of appropriate focusing solutions. In the third chapter, we defined focusing tasks in detail. In subsequent chapters, we developed, analyzed, and evaluated focusing solutions for the selected focusing tasks. Finally, the implementation of generic sampling in a commercial data mining system contributes to the focusing component of a knowledge discovery system.

In summary, this dissertation is the first comprehensive contribution to the development, analysis, and evaluation of focusing solutions for data mining.

7.2   More Related Work

We already described and analyzed relevant existing approaches towards focusing solutions for the selected focusing tasks in section 4.1. In this section, we briefly discuss more related work. This work includes contributions to solve other focusing tasks as well as technical aspects of sampling databases. In order to point interested readers to some of this work, we list some references in these areas but do not present full descriptions of these efforts.

Solution Approaches for Different Focusing Tasks

In this dissertation, we restricted our attention to focusing tuples on simple subsets or constrained subsets according to representativeness. We used specific filter and wrapper evaluation criteria to measure representativeness of focusing outputs. We considered classification goals and two different data mining algorithms, top down induction of decision trees and nearest neighbor classifiers. More related work also covers approaches that either deal with different focusing specifications or different focusing contexts:

- *Focusing Attributes*

  For example, many approaches consider the focusing input attributes rather than tuples. Blum & Langley (1997) as well as Aha & Bankert (1996) present overviews of *feature selection* approaches in machine learning such as the RELIEF (Kira & Rendell, 1992a; 1992b) and FOCUS (Almuallim & Dietterich, 1991) systems. From a statistics point of view, Richeldi & Rossotto (1997) report on combining statistical techniques and search heuristics to perform effective feature selection. Other approaches to solve focusing tasks with focusing input attributes include *peepholing* for mega-induction (Catlett, 1992), *greedy attribute selection* (Caruana & Freitag, 1994), and *attribute focusing* in software production process control (Bhandari, 1994).

- *Focusing for Association Rules*

  Although most existing efforts towards focusing solutions work within contexts of classification goals, Toivonen (1996) and Zaki *et al.* (1997) describe experiments on using random sampling for finding association rules. They propose to draw random samples from databases and to generate associations on these samples rather than on the entire database. These approaches reduce I/O overhead and the number of transactions that association rule algorithms need to consider in order to extract exact or approximate associations.

- *Active Learning*

  Another interesting line of research related to focusing is *active learning* (Cohn *et al.*, 1994; 1996). Active learning mainly deals with control tasks in robotics. The general idea is that learning systems actively try to select *optimal* next training examples in order to gain maximum progress during the learning process. In this sense, active learning paradigms use fewer examples since they avoid learning on examples that do not lead to learning progress.

Technical Aspects of Sampling Databases

From database technology perspectives, sampling databases received much attention in the past. All approaches to sampling databases deal with technical issues in realizing statistical sampling techniques in databases. Many of these efforts make use of specific data structures such as B^+ *trees, hash files*, and *spatial databases* that supply efficient implementations in database technologies. We refer to Olken (1993) and Olken & Rotem (1994) for surveys of sampling techniques in databases.

7.3   Future Work

In comparison to initial expectations at the beginning of this work on focusing solutions, we still see the following limitations of the development, analysis, and evaluation of focusing solutions for data mining:

- *Data Access*

  The current implementation of a unified approach to focusing solutions in generic sampling still works on flat files in main memory rather than databases. Although there is no restriction in the specification of algorithms which makes direct access to databases impossible, we have to select data from databases manually and to prepare the focusing input in separate transformation steps in data preparation, before we apply focusing solutions in generic sampling.

- *Execution Time*

  Occasionally, execution time of focusing solutions is still high, if we specify advanced options in generic sampling. For example, if the data set contains many quantitative attributes, computation of attribute relevance weights takes several minutes up to a few hours.

- *Focusing Advice*

  Focusing advice presented at the end of experimental studies consolidates experiences from experiments with ten different focusing solutions on eight selected databases from the UCI machine learning repository. Hence, this advice contains heuristics rather than theoretically proven properties of relations between specific focusing tasks and focusing success of focusing solutions. The only theoretical results present an average case analysis for specific classification goals and nearest neighbor classifiers in combination with simple random sampling.

If we take into account limitations of the work presented in this dissertation and consider ideas which we were not able to work on in detail, we raise the following issues for future work:

- *Enhancements to Implementations*

  First of all, we think of improvements to the current implementation of generic sampling. For example, we want to integrate more sophisticated similarity measures that utilize additional background knowledge as well as consider alternative discretization approaches.

- *Data Access*

  We propose to re-implement focusing solutions for data mining such that they supply direct access to databases and utilize advantages and capabilities of database technologies.

- *Focusing Advice*

  We also like to extend analytical studies in order to get more insights on relations between focusing tasks and focusing success of focusing solutions. However, the current experiences in analytical studies indicate that theoretical analyses of these relations are hard and not possible for general cases.

- *Experimental Studies*

  In the future, we will conduct more experiments on different data sets in order to manifest results of current experimental studies. Since we also want to stress relevance of focusing solutions in practice, we will perform more experiments on industrial databases as well.

7.4   Closing Remarks

Initially, we started research in the area of KDD and data mining since we wanted
to contribute to a field which is both scientifically interesting and practically
relevant. In particular, we focused attention on focusing challenges in KDD
since we recognized vast amounts of data collected, stored, and maintained, as
well as limitations of existing data mining algorithms to small data sets. All in
all, the decision to take the opportunity to work on a dissertation in an industrial
context was correct.

In conclusion, we perceived that theory and practice often diverge. From
an industrial point of view, research is only successful, if it is beneficial and
valuable in business terms. From this perspective, the true value of research and
technology is its applicability in practice.

As most dissertations do, we only solved small pieces in a whole puzzle. And
as we all know, a dissertation usually raises more open questions than it actually
solves existing challenges.

Bibliography

Adriaans, P., & Zantinge, D. (1996). *Data Mining*. Harlow, England: Addison-Wesley.

Agrawal, R., Imielinski, T., & Swami, A. (1993). Mining Association Rules between Sets of Items in Large Databases. in: Buneman, P., & Jajodia, S. (eds.). *Proceedings of the ACM SIGMOD International Conference on Management of Data*. May, 24-27, Minneapolis, Minnesota. SIGMOD Record 22 (2), pp. 207-216.

Aha, D.W., Kibler, D., & Albert, M.K. (1991). Instance-Based Learning Algorithms. *Machine Learning*, 6, p. 37-66.

Aha, D.W., & Bankert, R.L. (1996). A Comparative Evaluation of Sequential Feature Selection Algorithms. in: Fisher, D., & Lenz, H.-J. (eds.). *Learning from Data: AI and Statistics V*. New York, NY: Springer, pp. 199-206.

Al-Attar, A. (1997). CRITIKAL: Client-Server Rule Induction Technology for Industrial Knowledge Acquisition from Large Databases. in: Scheuerman, P. (ed.). *Proceedings of the 7th International Workshop on Research Issues in Data Engineering*. April, 7-8, Birmingham, England. Los Alamitos, CA: IEEE Computer Society Press, pp. 70-72.

Almuallim, H., & Dietterich, T.G. (1991). Learning with Many Irrelevant Features. in: *Proceedings of the 9th National Conference on Artificial Intelligence*. July, 14-19, Anaheim, CA. Cambridge, MA: MIT Press, pp. 547-552.

Bacher, J. (1994). *Clusteranalyse*. München: Oldenbourg Verlag.

Baim, P.W. (1988). A Method for Attribute Selection in Inductive Learning Systems. *IEEE Transactions on Pattern Analysis and Machine Intelligence*, Vol. 10, pp. 888-896.

Barreis, E.R. (1989). *Exemplar-Based Knowledge Acquisition*. Boston, MA: Academic Press.

Bhandari, I. (1994). Attribute Focusing: Machine-Assisted Knowledge Discovery Applied to Software Production Process Control. *Knowledge Acquisition*, Vol. 6, pp. 271-294.

T. Reinartz: Focusing Solutions for Data Mining, LNAI 1623, pp. 239-251, 1999
© Springer-Verlag Berlin Heidelberg 1999

Bentley, J.L. (1975). Multidimensional Binary Search Trees Used for Associative Searching. *Communications of the ACM*, Vol. 18, pp. 509-517.

Blum, A.L., & Langley, P. (1997). Selection of Relevant Features and Examples in Machine learning. *Artificial Intelligence*, 97, pp. 245-271.

Bock, H.H. (1974). *Automatische Klassifikation*. Göttingen: Vandenhoeck & Ruprecht.

Borgelt, C., Gebhardt, J., & Kruse, R. (1996). Concepts for Probabilistic and Possibilistic Induction of Decision Trees on Real World Data. in: Zimmermann, H.-J. (ed.) *Proceedings of the 4th European Congress on Intelligent Techniques and Soft Computing*. September, 2-5, Aachen, Germany. Aachen, Germany: Verlag Mainz, Wissenschaftsverlag, Vol. 3, pp. 1556-1560.

Börner, K., Faßauer, R., & Wode, H. (1996). Conceptual Analogy: Conceptual Clustering for Informed and Efficient Analogical Reasoning. in: Burkhardt, H.-D., & Lenz, M. (eds.). *Proceedings of the 4th German Workshop on Case-Based Reasoning*. March, 20-22, Berlin. Berlin: Springer, pp. 78-86.

Brachman, R.J., & Anand, T. (1994). The Process of Knowledge Discovery in Databases: A First Scetch. in: Fayyad, U.M., & Uthurasamy, R. (eds.). *Proceedings of KDD '94*. July, 31, - August, 1, Seattle, Washington. Menlo Park, CA: AAAI Press, pp. 1-11.

Brachman, R.J., & Anand, T. (1996). The Process of Knowledge Discovery in Databases: A Human-Centered Approach. in: Fayyad, U.M., Piatetsky-Shapiro, G., Smyth, P., & Uthurasamy, R. (eds.). *Advances in Knowledge Discovery and Data Mining*. Menlo Park, CA: AAAI Press, pp. 37-58.

Brachman, R.J., Khabaza, T., Kloesgen, W., Piatetsky-Shapiro, G., & Simoudis, E. (1996). Mining Business Databases. *Communications of the ACM*, Vol. 39, No. 11, pp. 42-48.

Breiman, L., Friedman, J. H., Olshen, A. & Stone, C. J. (1984). *Classification and Regression Trees*. Wadsworth: Belmont.

Breitner, C.A., Freyberg, A., & Schmidt, A. (1995). Towards Flexible and Integrated Environment for Knowledge Discovery. in: Ong, K., Conrad, S., & Ling, T.W. (eds.). *Knowledge Discovery and Temporal Reasoning in Deductive and Object-Oriented Databases: Proceedings of the DOOD '95 Post-Conference Workshops*. December, 4-8, Department of Information Systems and Computer Science, National University of Singapore, pp. 28-35.

Brodley, C. & Smyth, P. (1997). Applying Classification Algorithms in Practice. *Statistics and Computing*, 7, pp. 45-56.

Carbone, P.L., & Kerschberg, L. (1993). Intelligent Mediation in Active Knowledge Mining: Goals and General Description. in: Piatetsky-Shapiro, G. (ed.). *Proceedings of 1993 AAAI Workshop on Knowledge Discovery in Databases*. July, 11-12, Washington, D.C. Menlo Park, CA: AAAI Press, pp. 241-253.

Caruana, R., & Freitag, D. (1994). Greedy Attribute Selection. in: Cohen, W.W., & Hirsh, H. (eds.). *Proceedings of the 11th International Conference on Machine Learning*. July, 10-13, Rutgers University, New Brunswick, N.J. San Mateo, CA: Morgan Kaufmann, pp. 28-36.

Catlett, J. (1991). *Megainduction: Machine Learning on Very Large Databases.*. Ph.D. Thesis, University of Sydney, Australia.

Catlett, J. (1992). Peepholing: Choosing Attributes Efficiently for Megainduction. in: Sleeman, D., & Edwards, P. (eds.). *Proceedings of the 9th International Conference on Machine Learning*. July, 1-3, Aberdeen, Scotland. San Mateo, CA: Morgan Kaufmann, pp. 49-54.

Chang, C.-L. (1974). Finding Prototypes for Nearest Neighbor Classifiers. *IEEE Transactions on Computers*, Vol. 23, No. 11, pp. 1179-1184.

Chapman, P., Clinton, J., Hejlesen, J.H., Kerber, R., Reinartz, T., Shearer, C., & Wirth, R. (1998). *Initial Process Model*. Deliverable 2.1.1 of ESPRIT Project CRISP-DM.

Cochran, W.G. (1977). *Sampling Techniques*. New York: John Wiley & Sons.

Cohn, D., Atlas, L., & Ladner, R. (1994). Improving Generalization with Active Learning. *Machine Learning*, 5, pp. 201-221.

Cohn, D.A., Ghahramani, Z., & Jordan, M.I. (1996). Active Learning with Statistical Models. *Journal of Artificial Intelligence Research*, 4, pp. 129-145.

Cooper, G.F., & Herskovits, E. (1992). A Bayesian Method for the Induction of Probabilistic Networks from Data. *Machine Learning*, 9, pp. 309-347.

Cost, S., & Salzberg, S. (1993). A Weighted Nearest Neighbor Algorithm for Learning with Symbolic Features. *Machine Learning*, 10, pp. 37-78.

Darlington, J., Guo, Y., Sutiwaraphun, J., & To, H.W. (1997). Parallel Induction Algorithms for Data Mining. in: Liu, X., Cohen, P., & Berthold, M. (eds.). *Advances in Intelligent Data Analysis*. August, 4-6, London, UK. Berlin: Springer, pp. 437-445.

Dasarathy, B.V. (1991). *Nearest Neighbor (NN) Norms: NN pattern Classification Techniques.*. Los Alamitos, CA: IEEE Computer Society Press.

Datta, P., & Kibler, D. (1995). Learning Prototypical Concept Descriptions. in: Prieditis, A., & Russell, S. (eds.). *Proceedings of the 12th International Conference on Machine Learning.* July, 9-12, Tahoe City, CA. San Mateo, CA: Morgan Kaufmann, pp. 158-166.

Datta, P. (1997). Applying Clustering to the Classification Problem. in: *Proceedings of the 14th National Conference on Artificial Intelligence.* July, 27-31, Providence, Rhode Island. Menlo Park, CA: AAAI Press, pp. 826.

Datta, P., & Kibler, D. (1997). Symbolic Nearest Mean Classifiers. in: *Proceedings of the 14th National Conference on Artificial Intelligence.* July, 27-31, Providence, Rhode Island. Menlo Park, CA: AAAI Press, pp. 82-87.

Dempster, A.P., Laird, N.M., & Rubin, D.B. (1977). Maximum Likelihood from Incomplete Data via the EM Algorithm. *Journal of Royal Statistical Society,* Vol. 39, pp. 1-38.

Dougherty, J., Kohavi, R., & Sahami, M. (1995). Supervised and Unsupervised Discretization of Continuous Features. in: Prieditis, A., & Russell, S. (eds.). *Proceedings of the 12th International Conference on Machine Learning.* July, 9-12, Tahoe City, CA. Menlo Park, CA: Morgan Kaufmann, pp. 194-202.

Efron, B., & Tibshirani, R.J. (1993). *An Introduction to the Bootstrap.* Chapman & Hall, New York.

Engels, R. (1996). Planning Tasks for Knowledge Discovery in Databases: Performing Task-oriented User-guidance. in: Simoudis, E., Han, J., & Fayyad, U. (eds.). *Proceedings of the 2nd International Conference on Knowledge Discovery and Data Mining.* August, 2-4, Portland, Oregon. Menlo Park, CA: AAAI Press, pp. 170-175.

Ester, M., Kriegel, H.-P., & Xu, X. (1995). Knowledge Discovery in Large Spatial Databases: Focusing Techniques for Efficient Class Identification. in: Egenhofer, M.J., & Herring, J.R. (eds.). *Advances in Spatial Databases,* 4th International Symposium. August, 6-9, Portland, Maine, USA. New York, NY: Springer, pp. 67-82.

Fayyad, U.M. (1996). Data Mining and Knowledge Discovery: Making Sense out of Data. *IEEE Expert,* Vol. 11, No. 5, pp. 20-25.

Fayyad, U.M. (1997). Editorial in *Data Mining and Knowledge Discovery,* Vol. 1, No. 1.

Fayyad, U.M., & Irani, K. (1993). Multi-interval Discretization of Continuous-valued Attributes for Classification Learning. in: Bajcsy, R. (ed.). *Proceedings of the 13th International Joint Conference on Artificial Intelligence.* August, 28 - September, 3, Chambery, France. Los Alamitos, CA: IEEE Computer Society Press, pp. 1022-1027.

Fayyad, U.M., & Uthurasamy, R. (1994). *Proceedings of KDD '94*. July, 31, - August, 1, Seattle, Washington. Menlo Park, CA: AAAI Press.

Fayyad, U.M., Piatetsky-Shapiro, G., & Smyth, P. (1995). *Proceedings of the 1st International Conference on Knowledge Discovery in Databases*. August, 20-21, Montreal, Canada. Menlo Park, CA: AAAI Press.

Fayyad, U., Piatetsky-Shapiro, G., & Smyth, P. (1996). Knowledge Discovery and Data Mining: Towards a Unifying Framework. in: Simoudis, E., Han, J., & Fayyad, U. (eds.). *Proceedings of the 2nd International Conference on Knowledge Discovery and Data Mining*. August, 2-4, Portland, Oregon. Menlo Park, CA: AAAI Press, pp. 82-88.

Famili, A., Shen, W.-M., Weber, R., & Simoudis, E. (1997). Data Preprocessing and Intelligent Data Analysis. *Intelligent Data Analysis*, Vol. 1, No. 1, http://www.elsevier.com/locate/ida.

Fürnkranz, J. (1997). More Efficient Windowing. in: *Proceedings of the 14th National Conference on Artificial Intelligence*. July, 27-31, Providence, Rhode Island. Menlo Park, CA: AAAI Press, pp. 509-514.

Galal, G., Cook, D.J., & Holder, L.B. (1997). Improving Scalability in a Scientific Discovery System by Exploiting Parallelism. in: Heckerman, D., Mannila, H., Pregibon, D., & Uthurasamy, R. (eds.). *Proceedings of the 3rd International Conference on Knowledge Discovery and Data Mining*. August, 14-17, Newport Beach, CA. Menlo Park, CA: AAAI Press, pp. 171-174.

Genari, J.H. (1989). *A survey of clustering methods*. Technical Report 89-38, University of California, Irvine, CA.

Genari, J.H., Langley, P., & Fisher, D. (1989). Models of Incremental Concept Formation. *Artificial Intelligence*, 40, pp. 11-61.

Grimmer, U., & Mucha, A. (1997). Datensegmentierung mittels Clusteranalyse. in: Nakhaeizadeh, G. (ed.). *Data Mining: Theoretische Aspekte und Anwendungen*. Heidelberg: Physica Verlag, pp. 109-141.

Goos, K. (1995). *Fallbasiertes Klassifizieren: Methoden, Integration und Evaluation*. Dissertation, Universität Würzburg.

Hartigan, J.A. (1975). *Clustering Algorithms*. New York, NY: John Wiley & Sons, Inc.

Hartung, J., & Elpelt, B. (1986). *Multivariate Statistik*. München: Oldenbourg-Verlag.

Hartung, J., Elpelt, B., & Klösener, K.-H. (1987). *Statistik*. München: Oldenbourg-Verlag.

Heckerman, D., Mannila, H., Pregibon, D., & Uthurasamy, R. (1997). *Proceedings of the 3rd International Conference on Knowledge Discovery and Data Mining*. August, 14-17, Newport Beach, CA. Menlo Park, CA: AAAI Press.

Higashi, M., & Klir, G.J. (1982). Measures of Uncertainty and Information based on Possibility Distributions. *International Journal of General Systems*, Vol. 9, pp. 43-58.

Hirschberg, D.S., & Pazzani, M.J. (1991). *Average-Case Analysis of a k-CNF Learning Algorithm*. Technical Report 91-50, Department of Information and Computer Science, University of California at Irvine.

Hong, S.J. (1994). *Use of Contextual Information for Feature Ranking and Discretization*. Research Report, RC 19664 (87195), IBM Research, CA.

John, G.H. (1997). *Enhancements to the Data Mining Process*. Dissertation, Stanford University, California.

John, G.H., Kohavi, R., & Pfleger, K. (1994). Irrelevant Features and the Subset Selection Problem. in: Cohen, W.W., & Hirsh, H. (eds.). *Proceedings of the 11th International Conference on Machine Learning*. July, 10-13, Rutgers University, New Brunswick, N.J. San Mateo, CA: Morgan Kaufmann, pp. 121-129.

John, G.H., & Langley, P. (1996). Static Versus Dynamic Sampling for Data Mining. in: Simoudis, E., Han, J., & Fayyad, U. (eds.). *Proceedings of the 2nd International Conference on Knowledge Discovery and Data Mining*. August, 2-4, Portland, Oregon. Menlo Park, CA: AAAI Press, pp. 367-370.

Kauderer, H., & Nakhaeizadeh, G. (1997). The Effect of Alternative Scaling Approaches on the Performance of Different Supervised Learning Algorithms: An Empirical Study in the Case Credit Scoring. in: *Collected Papers from the 1997 Workshop on AI Approaches to Fraud Detection and Risk Management*. July, 27th, Providence, Rhode Island. Menlo Park, CA: AAAI Press, pp. 39-42.

Khabaza, T., & Shearer, C. (1995). Data Mining with CLEMENTINE. in: *Proceedings of the IEE Colloquium on Knowledge Discovery in Databases*. February, 1-2, London. IEE Digest, No. 1995/021(B), pp. 1-5.

Kira, K., & Rendell, L.A. (1992a). The Feature Selection Problem: Traditional Methods and a New Algorithm. in: Swartout, W.R. (ed.). *Proceedings of the 10th National Conference on Artificial Intelligence*. July, 12-16, San Jose, CA. Cambridge, MA: MIT Press, pp. 129-134.

Kira, K., & Rendell, L.A. (1992b). A Practical Approach to Feature Selection. in: Sleeman, D., & Edwards, P. (eds.). *Proceedings of the 9th International Workshop on Machine Learning*. July, 1-3, Aberdeen, Scotland. San Mateo, CA: Morgan Kaufmann, pp. 249-256.

Kivinen, J., & Mannila, H. (1994). The power of sampling in knowledge discovery. in: *Proceedings of the 13th Symposium on Principles of Database Systems*. May, 24-26, Minneapolis, MN. New York, NY: ACM Press, pp. 77-85.

Klir, G.J., & Mariano, M. (1987). On the Uniqueness of a Possibility Measure of Uncertainty and Information. *Fuzzy Sets and Systems*, Vol. 24, pp. 141-160.

Kohavi, R. (1995a). *Wrappers for Performance Enhancement and Oblivious Decision Graphs*. Dissertation, Stanford University, California.

Kohavi, R. (1995b). A Study of Cross-Validation and Bootstrap for Accuracy Estimation and Model Selection. in: Mellish, C.S. (ed.). *Proceedings of the 14th International Joint Conference on Artificial Intelligence*. August, 20-25, Montreal, Canada. San Mateo, CA: Morgan Kaufmann, pp. 1137-1143.

Kohavi, R., & Sommerfield, D. (1995). $\mathcal{MLC}++$: *A Machine Learning Library in C++*, Stanford University, CA, USA, http://robotics.stanford.edu/users/ronnyk/mlc.html.

Kohavi, R., Sommerfield, D., & Dougherty, J. (1996). *Data Mining Using $\mathcal{MLC}++$: A Machine Learning Library in C++*. http://robotics.stanford.edu/~ronnyk.

Kohavi, R., Langley, P., & Yun, Y. (1997). The Utility of Feature Weighting in Nearest-Neighbor Algorithms. Poster Presentation at the 9th European Conference on Machine Learning. April, 23-25, Prague, Czech Republique.

Krichevsky, R.E., & Trofimov, V.K. (1983). The Performance of Universal Coding. *IEEE Transactions on Information Theory*, Vol. 27, No. 2, pp. 199–207.

Kufrin, R. (1997). Generating C4.5 Production Rules in Parallel. in: *Proceedings of the 14th National Conference on Artificial Intelligence*. July, 27-31, Providence, Rhode Island. Menlo Park, CA: AAAI Press, pp. 565-570.

Lakshminarayan, K., Harp, S.A., Goldman, R., & Samad, T. (1996). Imputation of Missing Data Using Machine Learning Techniques. in: Simoudis, E., Han, J., & Fayyad, U. (eds.). *Proceedings of the 2nd International Conference on Knowledge Discovery and Data Mining*. August, 2-4, Portland, Oregon. Menlo Park, CA: AAAI Press, pp. 140-145.

Langley, P. (1996). *Elements of Machine Learning*. San Francisco, CA: Morgan Kaufmann.

Langley, P., Iba, W., & Thompson, K. (1992). An Analysis of Bayesian Classifiers. in: Swartout, W.R. (ed.). *Proceedings of the 10th National Conference on Artificial Intelligence*. July, 12-16, San Jose, CA. Cambridge, MA: MIT Press, pp. 223-228.

Langley, P., & Iba, W. (1993). Average-Case Analysis of a Nearest Neighbor Algorithm. in: Bajcsy, R. (ed.). *Proceedings of the 13th International Joint Conference on Artificial Intelligence*. August, 28 - September, 3, Chambery, France. Los Alamitos, CA: IEEE Computer Society Press, pp. 889-894.

Langley, P. & Simon, H. (1995). Applications of Machine Learning and Rule Induction. *Communications of the ACM*, Vol. 38, No. 11, pp. 55-63.

Li, B. (1997). *PC45: A Parallel Version of C4.5 Built with Persistent Linda System*. http://info.gte.com/~kdd/sift/pc45.html.

Linde, Y., Buzo, A., & Gray, R. (1980). An Algorithm for Vector Quantizer Design. *IEEE Transactions on Communications*, 28, pp. 85-95.

Little, R.J., & Rubin, D.B. (1987). *Statistical Analysis with Missing Data*. New York: John Wiley & Sons.

Lockemann, P.C., & Schmidt, J.W. (1987). *Datenbank-Handbuch*. Berlin: Springer.

Mackin, N. (1997). Application of WhiteCross MPP Servers to Data Mining. in: *Proceedings of the 1st International Conference on the Practical Application of Knowledge Discovery and Data Mining*. April, 23-25, London, UK. London, UK: Practical Application Company Ltd., pp. 1-8.

Mantaras de, L. (1991). A Distance-based Attribute Selection Measure for Decision Tree Induction. *Machine Learning*, 6, pp. 81-92.

Mason, R.L., Gunst, R.F., & Hess, J.L. (1989). *Statistical Design and Analysis of Experiments*. New York, NY: John Wiley & Sons.

Matheus, C.J., Chan, P.K., & Piatetsky-Shapiro, G. (1993). Systems for Knowledge Discovery in Databases. *IEEE Transactions on Knowledge and Data Engineering*, Vol. 5, No. 6, pp. 903-913.

McLaren, I., Babb, E., & Bocca, J. (1997). DAFS: Data Mining File Server. in: Scheuerman, P. (ed.). *Proceedings of the 7th International Workshop on Research Issues in Data Engineering*. April, 7-8, Birmingham, England. Los Alamitos, CA: IEEE Computer Society Press, pp. 62-65.

Michie, D. Spiegelhalter, D., & Taylor, C. (1994). *Machine Learning, Neural and Statistical Classification*. London, UK: Ellis-Horwood-Series in Artificial Intelligence.

Mitchel, T.M. (1997). *Machine Learning*. Boston, MA: McGraw-Hill.

Moller, M. (1993). Supervised Learning on Large Redundant Training Sets. *International Journal of Neural Systems*, Vol. 4, No. 1, pp. 15-25.

Mucha, H.-J. (1992). *Clusteranalyse mit Mikrocomputern*. Berlin: Akademie Verlag.

Murphy, P.M., & Aha, D. (1994). *UCI Repository of Machine Learning Databases.* ftp://ics.uci.edu/pub/machine-learning-databases.

Nakhaeizadeh, G., & Schnabl, A. (1997). Development of Multi-Criteria Metrics for Evaluation of Data Mining Algorithms. in: Heckerman, D., Mannila, H., Pregibon, D., & Uthurasamy, R. (eds.). *Proceedings of the 3rd International Conference on Knowledge Discovery and Data Mining.* August, 14-17, Newport Beach, CA. Menlo Park, CA: AAAI Press, pp. 37-42.

Nakhaeizadeh, G., Reinartz, T., & Wirth, R. (1997). Wissensentdeckung in Datenbanken und Data Mining: Ein Überblick. in: Nakhaeizadeh, G. (ed.). *Data Mining: Theoretische Aspekte und Anwendungen.* Heidelberg: Physica Verlag, pp. 1-33.

Ng, R.T., & Han, J. (1994). Efficient and Effective Clustering Methods for Spatial Data Mining. in: Bocca, J.B., Jarke, M., & Zaniolo, C. (eds.). *Proceedings of the 20th International Conference on Very Large Databases.* September, 12-15, Santiago de Chile, Chile. San Mateo, CA: Morgan Kaufmann, pp. 144-155.

Olken, F. (1993). *Random Sampling from Databases.* Dissertation, University of California at Berkeley, CA, USA.

Olken, F., & Rotem, D. (1994). *Random Sampling from Databases - A Survey.* Draft Manuscript, University of California at Berkeley, CA, USA.

Parsa, I. (1997). A Knowledge Discovery and Data Mining Tools Competition. in: Piatetsky-Shapiro, G. (ed.). *Knowledge Discovery Nuggets*, No. 19.

Parzen, E. (1960). *Modern Probability Theory and its Applications.* New York: John Wiley & Sons, Inc.

Pearl, J. (1988). *Probabilistic Reasoning in Intelligent Systems.* San Mateo, CA: Morgan Kaufman.

Piatetsky-Shapiro, G. (1991). *Proceedings of 1991 AAAI Workshop on Knowledge Discovery in Databases.* July 14-15, Anaheim, CA. Menlo Park, CA: AAAI Press.

Piatetsky-Shapiro, G. (1993). *Proceedings of 1993 AAAI Workshop on Knowledge Discovery in Databases.* July, 11-12, Washington, D.C. Menlo Park, CA: AAAI Press.

Provost, F., & Fawcett, T. (1997). Analysis and Visualization of Classifier Performance with Nonuniform Class and Cost Distributions. in: *Collected Papers from the 1997 Workshop on AI Approaches to Fraud Detection and Risk Management.* July, 27th, Providence, Rhode Island. Menlo Park, CA: AAAI Press, pp. 57-63.

Provost, F., & Kolluri, V. (1997). Scaling up Inductive Algorithms: An Overview. in: Heckerman, D., Mannila, H., Pregibon, D. & Uthurasamy, R. (eds.). *Proceedings of the 3rd International Conference on Knowledge Discovery and Data Mining.* August, 14-17, Newport Beach, CA. Menlo Park, CA: AAAI Press, pp. 239-242.

Quinlan, J.R. (1986). Induction of Decision Trees. *Machine Learning*, 4, pp. 81-106.

Quinlan, J.R. (1993). *C4.5: Programs for Machine Learning.* San Mateo, CA: Morgan Kaufmann.

Reinartz, T. (1997). Advanced Leader Sampling for Data Mining Algorithms. in: Kitsos, C.P. (ed.). *Proceedings of the ISI '97 Satellite Conference on Industrial Statistics: Aims and Computational Aspects.* August, 16-17, Athens, Greece. Athens, Greece: University of Economics and Business, Department of Statistics, pp. 137-139.

Reinartz, T. (1998). Similarity-Driven Sampling for Data Mining. in: Zytkow, J.M. & Quafafou, M. (eds.). *Principles of Data Mining and Knowledge Discovery: Second European Symposium, PKDD '98.* September, 23-26, Nantes, France. Heidelberg: Springer, pp. 423-431.

Reinartz, T., & Wirth, R. (1995). The Need for a Task Model for Knowledge Discovery in Databases. in: Kodratoff, Y., Nakhaeizadeh, G., & Taylor, C. (eds.). Workshop notes *Statistics, Machine Learning, and Knowledge Discovery in Databases.* MLNet Familiarization Workshop, April, 28-29, Heraklion, Crete, pp. 19-24.

Richeldi, M., & Rossotto, M. (1997). Combining Statistical Techniques and Search Heuristics to Perform Effective Feature Selection. in: Nakhaeizadeh, G., & Taylor, C.C. (eds.). *Machine Learning and Statistics: The Interface.* New York, NY: John Wiley & Sons, pp. 269-291.

Rissanen, J. (1995). Stochastic Complexity. *Journal of the Royal Statistical Society* , Vol. 49, pp. 223-239.

Scheaffer, R.L., Mendenhall, W., & Ott, R.L. (1996). *Elementary Survey Sampling*, 5th Edition. New York, NY: Duxbury Press.

Sen, S., & Knight, L. (1995). A Genetic Prototype Learner. in: Mellish, C.S. (ed.). *Proceedings of the 14th International Joint Conference on Artificial Intelligence.* August, 20-25, Montreal, Quebec, Canada. San Mateo, CA: Morgan Kaufmann, Vol. I, pp. 725-731.

Simoudis, E., Han, J. & Fayyad, U. (1996). *Proceedings of the 2nd International Conference on Knowledge Discovery and Data Mining.* August, 2-4, Portland, Oregon. Menlo Park, CA: AAAI Press.

Simoudis, E. (1997). *Industry Applications of Data Mining: Opportunities and Challenges.* Presentation at the 1st International Conference on Principles of Knowledge Discovery in Databases. June, 25-27, Trondheim, Norway,

Skalak, D.B. (1993). Using a Genetic Algorithm to Learn Prototypes for Case Retrieval and Classification. in: Leake, D. (ed.). *Proceedings of the AAAI '93 Case-based Reasoning Workshop.* July, 11-15, Washington, DC. Menlo Park, CA: American Association for Artificial Intelligence, Technical Report WS-93-01, pp. 64-69.

Skalak, D.B. (1994). Prototype and Feature Selection by Sampling and Random Mutation Hill Climbing Algorithms. in: Cohen, W.W., & Hirsh, H. (eds.). *Proceedings of the 11th International Conference on Machine Learning.* July, 10-13, Rutgers University, New Brunswick, N.J. San Mateo, CA: Morgan Kaufmann, pp. 293-301.

Smyth, B., & Keane, M.T. (1995). Remembering to Forget. in: Mellish, C.S. (ed.). *Proceedings of the 14th International Joint Conference on Artificial Intelligence.* August, 20-25, Montreal, Quebec, Canada. San Mateo, CA: Morgan Kaufmann, Vol. I, pp. 377-382.

Stolorz, P., Nakamura, H., Mesrobian, E., Muntz, R.R., Shek, E.C., Mechoso, C.R., & Farrara, J.D. (1995). Fast Spatiotemporal Data Mining of Large Geophysical Datasets. in: Fayyad, U.M., Piatetsky-Shapiro, G., & Smyth, P. (eds.). *Proceedings of the 1st International Conference on Knowledge Discovery in Databases.* August, 20-21, Montreal, Canada. Menlo Park, CA: AAAI Press, pp. 300-305.

Taylor, C.C., & Nakhaeizadeh, G. (1997). Learning in Dynamically Changing Domains: Theory Revision and Context Dependence Issues. in: van Someren, M., & Widmer, G. (eds.). *Proceedings of 9th European Conference on Machine Learning.* April, 23-25, Prague, Czech Republique. Heidelberg: Springer, pp. 353-360.

Toivonen, H. (1996). Sampling Large Databases for Finding Association Rules. in: Vijayaraman, T.M., Buchmann, A.P., Mohan, C., & Sarda, N.L. (eds.). *Proceedings of the 22nd International Conference on Very Large Databases.* September, 3-6, Mumbai, India. San Mateo, CA: Morgan Kaufmann, pp. 134-145.

Tversky, A. (1977). Features of Similarity. *Psychological Review,* 84, pp. 327-352.

Ullman, J.D. (1982). *Principles of Database Systems.* Rockville, MD: Computer Science Press.

Valiant, L.G. (1984). A Theory of the Learnable. *Communications of the ACM,* Vol. 27, No. 11, pp. 1134-1142.

Weiss, S.M., & Indurkhya, N. (1997). *Predictive Data Mining.* San Francisco, CA: Morgan Kaufman.

Weiss, S.M., & Kulikowski, C.A. (1991). *Computer systems that Learn*. San Francisco, CA: Morgan Kaufmann.

Wess, S. (1995). *Fallbasiertes Problemlösen in wissensbasierten Systemen zur Entscheidungsunterstützung und Diagnostik: Grundlagen, Systeme und Anwendungen*. Dissertation, Universität Kaiserslautern.

Wess, S., Althoff, K.-D., & Derwand, G. (1994). Using kd-Trees to Improve the Retrieval Step in Case-Based Reasoning. in: Wess, S., Althoff, K.-D., & Richter, M.M. (eds.). *Topics in Case-Based Reasoning - Selected Papers from the 1st European Workshop on Case-Based Reasoning*. Berlin: Springer, pp. 167-181.

Wettschereck, D. (1994). *A Study of Distance-Based Machine Learning Algorithms*. Dissertation, Oregon State University, OR.

Wettschereck, D., Aha, D., & Mohri, T. (1995). *A Review and Comparative Evaluation of Feature Weighting Methods for Lazy Learning Algorithms*. Technical Report AIC-95-012, Naval Research Laboratory, Navy Center for Applied Research in Artificial Intelligence, Washington, D.C.

Wilson, D.R., & Martinez, T.R. (1997a). Instance Pruning Techniques. in: Fisher, D.H. (ed.). *Proceedings of the 14th International Conference on Machine Learning*. July, 8-12, Nashville, Tennessee. San Francisco, CA: Morgan Kaufmann, pp. 403-411.

Wilson, D.R. & Martinez, T.R. (1997b). Improved Heterogeneous Distance Functions. *Journal of Artificial Intelligence Research*, 6, pp. 1-34.

Wirth, J., & Catlett, J. (1988). Experiments on the Costs and Benefits of Windowing in ID3. in: Laird, J. (ed.). *Proceedings of the 5th International Conference on Machine Learning*. June, 12-14, University of Michigan, Ann Arbor. San Mateo, CA: Morgan Kaufmann, pp. 87-99.

Wirth, R., & Reinartz, T. (1995). *Towards a Task Model for Knowledge Discovery in Databases*. Report, Daimler-Benz AG, Ulm, Germany.

Wirth, R., & Reinartz, T. (1996). Detecting Early Indicator Cars in an Automotive Database: A Multi-Strategy Approach. in: Simoudis, E., Han, J., & Fayyad, U. (eds.). *Proceedings of the 2nd International Conference on Knowledge Discovery and Data Mining*. August, 2-4, Portland, Oregon. Menlo Park, CA: AAAI Press, pp. 76-81.

Wirth, R., Shearer, C., Grimmer, U., Reinartz, T., Schlösser, J., Breitner, C., Engels, R., & Lindner, G. (1997). Towards Process-Oriented Tool Support for KDD. in: Komorowski, J., & Zytkow, J. (eds.). *Principles of Data Mining and Knowledge Discovery: First European Symposium, PKDD '97*. June, 25-27, Trondheim, Norway. Heidelberg: Springer, pp. 243-253.

Zaki, M.J., Parthasarathy, S., Li, W., & Ogihara, M. (1997). Evaluation of Sampling for Data Mining of Association Rules. in: Scheuermann, P. (ed.). *Proceedings of the 7th Workshop on Research Issues in Data Engineering.* April, 7-8, Birmingham, England. Los Alamitos, CA: IEEE Computer Society Press, pp. 42-50.

Zhang, J. (1992). Selecting Typical Instances in Instance-Based Learning. in: Sleeman, D., & Edwards, P. (eds.). *Proceedings of the 9th International Conference on Machine Learning.* July, 1-3, Aberdeen, Scotland. San Mateo, CA: Morgan Kaufmann, pp. 470-479.

Zhou, X., & Dillon, T.S. (1991). A Statistical-heuristic Feature Selection Criterion for Decision Tree Induction. *IEEE Transactions on Pattern Analysis and Machine Intelligence*, Vol. 13, pp. 834-841.

Acknowledgments

At the end of my studies in computer science and mathematics in summer 1994, I met Wolfgang Hanika for the first time, and I asked him about possibilities to work as a Ph.D. student at Daimler-Benz. His answer was "Fine, no problem.", and I thought, he was kidding, but he was not. I went to Prof. Michael M. Richter and checked with him whether he was willing to be my supervisor, if I do my dissertation at Daimler-Benz. He agreed and immediately phoned Prof. Gholamreza Nakhaeizadeh who was (and still is) head of the machine learning group at Daimler-Benz, Research & Technology, in Ulm. Prof. Michael M. Richter explained my situation and recommended me to Prof. Gholamreza Nakhaeizadeh. Then, I applied for a Ph.D. position at Daimler-Benz, went to an interview, and finally Daimler-Benz offered me a three years contract. My life as a Ph.D. student in industry was born.

I thank Wolfgang Hanika, Prof. Michael M. Richter, and Prof. Gholamreza Nakhaeizadeh for their support and initiatives. Wolfgang Hanika was head of the department I was (and still am) working in, Prof. Michael M. Richter became my first supervisor, and Prof. Gholamreza Nakhaeizadeh my second. All of them contributed to the success of my dissertation.

Wolfgang Hanika provided all organizational support, freedom to work independently, and opportunities to travel to workshops and conferences. In particular, Daimler-Benz allowed me to stay as a visiting scholar at Stanford University for six months in 1996, and I thank Paul Mehring at the Daimler-Benz research center in Palo Alto for office space and machine facilities. Prof. Michael M. Richter always found time to discuss my work and encouraged me to make progress as well as to finish in the end. Similarly, I also had a lot of fruitful discussions with Prof. Gholamreza Nakhaeizadeh. He was (and still is) a pleasant boss and a competent researcher in knowledge discovery and data mining.

At Stanford, I had the chance to present my developments in seminars at Stanford University, University of California at Berkeley, as well as the Jet Propulsion Laboratory in Pasadena. I thank Pat Langley, Larry Mazlack, and Lars Asker for organizing these talks. During my presentations and in separate sessions, I discussed my work with Leo Breiman, Jerome Friedman, Robert Gray, Trevor Hastie, Daphne Koller, Art Owen, Lofti Zadeh, and others. All of them

T. Reinartz: Focusing Solutions for Data Mining, LNAI 1623, pp. 253-255, 1999
© Springer-Verlag Berlin Heidelberg 1999

contributed at least one important aspect or one relevant reference, which was always helpful to make progress. My specific gratitude to Pat Langley who initiated a weekly discussion forum with George John and Mehran Sahami during my stay at Stanford. We also worked together on the average case analysis, and it was him who encouraged me to search for the correct formulas until I found them.

I further thank Harald Kauderer and Rüdiger Wirth, two colleagues at Daimler-Benz. Both read initial versions of this dissertation and gave valuable comments to improve this document. Moreover, Harald and Rüdiger became more than colleagues over time, and I enjoyed the time that we spent (and still spend) together. In a similar way, I thank all past and current colleagues in our department for the positive atmosphere at Daimler-Benz: Sigfried Bell, Hannah Blau, Christian Borgelt, Karl Dübon, Mehmet Göker, Udo Grimmer, Stefan Ohl, Folke Rauscher, Jutta Stehr, Elmar Steurer, and Alexander Vierlinger.

Since this dissertation and its contents were developed as part of the innovative project "Data Mining with Machine Learning", I also thank all past and current colleagues in this project, especially Christoph Breitner, Peter Brockhausen, Robert Engels, Guido Lindner, and Jörg Schlösser. They are all Ph.D. students, and we often discussed our ideas and enjoyed the social events.

Furthermore, I thank Christian Borgelt who implemented the initial data structures at Daimler-Benz, which I reused and adapted for my own programs, and contributed to the average case analysis. Thomas Klenk implemented the discretization approaches, and Ralf Steinwender realized the hypothesis tests for filter evaluation as well as all the scripts that were necessary to perform the experimental study. ISL provided their data mining system CLEMENTINE and its source code, and Julian Clinton and Colin Shearer gave tips and hints in Poplog for the integration of generic sampling into CLEMENTINE. Julian Clinton also read the final version of this dissertation and pointed out a few remaining phrasing problems. For further technical support, I thank Udo Grimmer and Martin Hahn.

I also want to mention some friends who I met in school, during my study, or later. I certainly cannot list all people who were part of my life in more or less important roles and during more or less long periods, but I wish to emphasize those few friends who are still a great support in all situations, either private or business-related: Lothar Becker, Markus Bücker, Christoph Globig, Andre Hüttemeister, Stefan Müller, Thomas Scherer, and Stefan Simon.

Last but definitely not least, I owe the most important gratitude my family. First of all, my parents, Renate and Adolf Reinartz, supported me and my own family whenever we needed help. I particularly never forget what they did for us during Ines' (first) pregnancy while I unfortunately was at Stanford. Without their support, this dissertation and all its prerequisites would have been not possible.

Similarly, my sister Hella Antheck and her family were of great help throughout all the years of my studies. In particular, they helped in taking care of our son Lukas while Ines was in hospital and I was busy preparing my oral's defense.

Finally, it is my specific concern to thank my wife Ines and our son Lukas. Both spent a hard time with me, and often I did not have enough time for them. Ines did most of the work at home, and she took our son and spent some time at my parents or at some friends several times in order to help me in finding spare time to work on my dissertation. Whenever I came home late in the evening and needed someone to tell my successes or failures, she was there and helped me to change my mood at the end of less successful days. Likewise, Lukas often managed to enable my recovery after an exhaustive long week by just smiling and laughing, "talking" to me, or playing with the ball and me. Without their love and patience, I would never have found the motivation to finish this dissertation.

Appendix A

Notations

In this appendix, we list notations for indices, variables, and functions, as well as for algorithms and procedures. For each entry, we remind the specific notation, present a brief description of its meaning, provide reference numbers that refer to definitions, propositions, algorithms, or procedures which introduce or use this particular notation, and also specify page references where to find these definitions, propositions, algorithms, or procedures.

A.1 Indices, Variables, and Functions

Table A.1 lists all notations for indices, variables, and functions.

Table A.1: Notations for Indices, Variables, and Functions

Notation	Description	Reference	Page
$\tilde{a}$	estimated NN accuracy for WSCs	5.1.2	164
$\hat{a}$	expected average NN accuracy for WSCs	5.1.4, 5.1.5	167, 169
A	$= \{a_1, \ldots, a_j, \ldots, a_N\}$, set of attributes	2.4.2	26
A	powerset of A	2.4.2	26
$A_{[j;\,j']}$	$= \{a_j, a_{j+1}, \ldots, a_{j'-1}, a_{j'}\}$, projection of attributes	2.4.4	27

T. Reinartz: Focusing Solutions for Data Mining, LNAI 1623, pp. 257-265, 1999
© Springer-Verlag Berlin Heidelberg 1999

Notation	Description	Reference	Page
a_j	jth attribute in A	2.4.1	25
a_{jk}	kth value of attribute a_j	2.4.1	25
a_j	unknown value of attribute a_j	2.5.10	42
(A, T)	(database) table	2.4.2	26
$a(\phi, T_{test})$	accuracy of ϕ on T_{test}	3.4.9	68
α	significance level	3.4.2	60
b_l	begin of lth interval	4.2.6	128
C	clustering	4.1.3	95
C_R	set of represented clusters	5.1.1	162
c_l	lth cluster	4.1.3	95
$C(A, T)$	set of data characteristics	3.3.1	52
d	distance measure	C.0.1	277
D	data mining algorithm	3.3.1	52
δ	similarity threshold	4.1.4	100
$d(\alpha, T)$	expected distribution	3.4.2	60
$d_{D_j}(\alpha, T)$	expected distribution for distributions of attribute a_j	3.4.5	63
$d_J(\alpha, T)$	expected distribution for joint distributions	3.4.6	64
$d_{S_j}(\alpha, T)$	expected distribution for means of attribute a_j	3.4.3	62
$d_{V_j}(\alpha, T)$	expected distribution for variances of attribute a_j	3.4.4	62
$dom(a_j)$	$= \{a_{j1}, \ldots, a_{jk}, \ldots, a_{jN_j}\}$, domain of attribute a_j	2.4.1	25
E	evaluation criterion	3.5.2	84
E_{D_j}	isolated evaluation criterion for distributions	B.1.2	268
E_J	isolated evaluation criterion for joint distributions	B.1.3	269
E_{S_j}	isolated evaluation criterion for modus and means	3.4.11	71
E_{V_j}	isolated evaluation criterion for variances	B.1.1	267
E_T	isolated evaluation criterion for execution time	3.4.15	78

Notation	Description	Reference	Page
E_M	isolated evaluation criterion for storage requirements	B.2.1	272
E_A	isolated evaluation criterion for accuracy	B.2.2	273
E_C	isolated evaluation criterion for complexity	B.2.3	273
E_{D_j}	comparative evaluation criterion for distributions	B.1.5	271
E_J	comparative evaluation criterion for joint distributions	B.1.6	271
E_{S_j}	comparative evaluation criterion for modus and means	3.4.12	75
E_{V_j}	comparative evaluation criterion for variances	B.1.4	270
E_T	comparative evaluation criterion for execution time	3.4.16	80
E_M	comparative evaluation criterion for storage requirements	B.2.4	274
E_A	comparative evaluation criterion for accuracy	B.2.5	274
E_C	comparative evaluation criterion for complexity	B.2.6	275
E_F	isolated cumulated filter evaluation criterion	3.4.13	76
E_F	comparative cumulated filter evaluation criterion	B.1.7	272
E_W	isolated cumulated wrapper evaluation criterion	3.4.17	82
E_W	comparative cumulated wrapper evaluation criterion	B.2.7	275
E_F	cumulated filter evaluation criterion	3.4.14	77
E_W	cumulated wrapper evaluation criterion	3.4.18	82
e_l	end of lth interval	4.2.6	128
$e(\phi, T_{test})$	error rate of ϕ on T_{test}	3.4.9	68

Notation	Description	Reference	Page
$f(f_{in}, f_{out})$	focusing criterion	3.2.4	50
F	focusing solution	3.5.2	84
f_{in}	focusing input	3.2.2	48
f_{out}	focusing output	3.2.3	50
F_{out}	set of focusing outputs	e.g., 3.4.11	e.g., 71
G	data mining goal	3.3.1	52
γ	parameter in Mucha's similarity threshold estimation	4.2.7	140
H_0	null hypothesis	3.4.1	59
H_1	alternative hypothesis	3.4.1	59
$H(T)$	entropy of T	2.5.2	35
$H_j(T)$	entropy of an attribute split on attribute a_j	2.5.3	35
$H_j^+(T)$	information gain of an attribute split on attribute a_j	2.5.4	36
$Het_1(T, T)$	heterogeneity between T and T, 1st alternative	4.1.4	96
$Het_2(T, T)$	heterogeneity between T and T, 2nd alternative	4.1.4	96
$Hom_1(T)$	homogeneity within T, 1st alternative	4.1.4	96
$Hom_2(T)$	homogeneity within T, 2nd alternative	4.1.4	96
i	tuple index	2.4.1	25
I	set of intervals	4.2.6	128
I_l	lth interval	4.2.6	128
$I_j^+(T)$	information gain ratio of an attribute split on attribute a_j	2.5.5	36
j	attribute index	2.4.1	25
$j_{ii'}$	first j which meets the first condition of (ii) in definition of $\succ$	4.2.2	117
k	value index	2.4.1	25
l	cluster / stratum / interval index	4.1.3, 4.2.5, 4.2.6	95, 125, 128
l_i	jth leader	4.1.4	100
L	number of clusters / strata / intervals	4.1.3, 4.2.5, 4.2.6	95, 125, 128

Notation	Description	Reference	Page
L_j	number of strata / intervals for attribute a_j	4.2.5, 4.2.6	125, 128
L_{N1}	number of clusters with class label a_{N1}	5.1.3	165
L_R	number of represented clusters	5.1.3	165
$l(\phi)$	number of levels in ϕ	3.4.10	69
M	number of tuples in T	2.4.2	26
M_R	number of represented tuples in T_R	5.1.1	162
m	number of tuples in focusing output	e.g., 4.1.1	e.g., 88
$m_j(T)$	modus of attribute a_j in T	2.4.7	30
m_l	focusing output size proportion for lth stratum	4.1.3	90
$m(D,T)$	storage requirements of D on T	3.4.8	67
$m(F,T)$	storage requirements of F on T	3.4.8	67
$\mu_j(T)$	mean of attribute a_j in T	2.4.8	30
N	number of attributes in A	2.4.2	26
N_j	number of values in $dom(a_j)$	2.4.1	25
$n_{jk}(T)$	number of tuples in T with $t_{ij} \equiv a_{jk}$	2.4.6	29
$n_{k_1...k_j...k_N}(T)$	number of tuples in T with $t_{ij} \equiv a_{jk_j} \forall j, 1 \le j \le N$	2.4.6	29
$n(\phi)$	number of nodes in ϕ	3.4.10	69
ϕ	classifier	2.5.1	32
p	probability function	3.2.1	51
P	predicate	3.2.3	50
$P(f_{in}, f_{out})$	prototype selection or prototype construction	4.1.8, 4.1.9	106, 108
Δ_1	threshold in PCE criterion	4.1.2	94
Δ_2	threshold in PCE criterion	4.1.2	94
$q_{D_j}(\alpha, f_{in}, f_{out})$	$= s_{D_j}(f_{in}, f_{out}) / d_{D_j}(\alpha, f_{out})$	3.4.5	63
$q_J(\alpha, f_{in}, f_{out})$	$= s_J(f_{in}, f_{out}) / d_J(\alpha, f_{out})$	3.4.6	64
$q_{S_j}(\alpha, f_{in}, f_{out})$	$= s_{S_j}(f_{in}, f_{out}) / d_{S_j}(\alpha, f_{out})$	3.4.3	62

Notation	Description	Reference	Page
$q_{V_j}(\alpha, f_{in}, f_{out})$	$= s_{V_j}(f_{in}, f_{out})/$ $d_{V_j}(\alpha, f_{out})$	3.4.4	62
$r_j(T)$	range of attribute a_j in T	2.4.9	30
R_{N1}	number of represented clusters with class label a_{N1}	5.1.3	165
$R(T)$	redundancy (factor) of T	2.4.11	31
S	set of strata	4.2.5	125
S_j	set of strata for attribute a_j	4.2.1	123
s_l	lth stratum	4.2.5	125
s_{jk}	kth stratum for qualitative attribute a_j	4.2.3	121
s_{jl}	lth stratum for quantitative attribute a_j	4.2.4	121
$s(f_{in}, f_{out})$	test statistic	3.4.2	60
$s_{D_j}(f_{in}, f_{out})$	test statistic for distribution of attribute a_j	3.4.5	63
$s_J(f_{in}, f_{out})$	test statistic for joint distribution	3.4.6	64
$s_{S_j}(f_{in}, f_{out})$	test statistic for means of attribute a_j	3.4.3	62
$s_{V_j}(f_{in}, f_{out})$	test statistic for variances of attribute a_j	3.4.4	62
$start$	start position for systematic sampling	4.1.2	89
$step$	step size for systematic sampling	4.1.2	89
sim	similarity measure	2.5.6	39
Sim	weighted cumulated similarity measure	2.5.9	40
$\mid Sim \mid$	number of similarities	4.2.7	139
sim_j	local similarity measure for attribute a_j	2.5.7, 2.5.8	39, 40
$\mid sim_j \mid$	number of local similarities for attribute a_j	4.2.6	139
$\sigma(x)$	standard deviation of x	e.g., 3.4.11	e.g., 71
$\sigma_j^2(T)$	variance of attribute a_j in T	2.4.10	30
$\sigma_j(T)$	standard deviation of attribute a_j in T	2.4.10	30
T	$= \{t_1, \ldots, t_i, \ldots, t_M\}$ set of tuples	2.4.2	26

Notation	Description	Reference	Page
T	powerset of T	2.4.2	26
T_R	set of represented tuples	5.1.2	164
T_{train}	training set	2.5.1	32
T_{test}	test set	2.5.1	32
$T_{[j;\,j']}$	$= \{(t_{ij}, t_{i(j+1)}, \ldots,$ $t_{i(j'-1)}, t_{ij'}) \mid t_i \in T\}$ projection of tuples	2.4.4	27
T_{jk}	$= \{t_i \in T \mid t_{ij} \equiv a_{jk}\}$ set of tuples with equal attribute values $a_{jk} \in$ $dom(a_j)$	2.5.3	35
$T_{\{\}}$	$= \{t_i \in T \mid \forall i,$ $1 \le i \le i-1: t_{i'} \ne t_i\}$ set of tuples without duplicates	2.4.11	31
$T(t_i)$	$= \{t_{i'} \in f_{in} \mid t_i \text{ is leader}$ of $t_{i'}\}$ set of tuples represented by leader t_i	4.2.8	142
t_i	ith tuple	2.4.1	25
t_{ij}	value of jth attribute of ith tuple	2.4.1	25
$t_{i[j;\,j']}$	$(t_{ij}, t_{i(j+1)}, \ldots,$ $t_{i(j'-1)}, t_{ij'})$ projection of tuple	2.4.4	27
$t(D, T)$	execution time of D on T	3.4.7	66
$t(F, T)$	execution time of F on T	3.4.7	66
θ	(population) parameter	3.4.1	59
Θ_0	subset of (population) parameter domain	3.4.1	59
Θ_0^c	complement of Θ_0	3.4.1	59
W	set of weights	4.3.1	152
w	sum of weights	2.5.9	40
w_i	prototype weight of tuple t_i	4.2.8	146
$width$	interval width in equal-width discretization	4.2.2	130
w_j	relevance weight of attribute a_j	2.5.9, 4.2.9	40, 149
x	object	2.5.6	39
X	set of objects	2.5.6	39

Notation	Description	Reference	Page
w_j	relevance weight of attribute a_j	2.5.9, 4.2.9	40, 149
$|\;|$	number of objects, absolute value, meet condition	2.4.3	26
$\lfloor x \rfloor$	largest integer value smaller than x	4.1.2	89
$\perp$	unknown value	2.4.1	27
$\succ$	order relation	4.2.2	117

A.2   Algorithms and Procedures

Table A.2 lists all notations for algorithms and procedures.

Table A.2: Notations for Algorithms and Procedures

Algorithm	Description	Reference	Page
ALESAM	advanced leader sampling	4.2.1	114
CLARANS	Clustering Large Applications based on RANdomized Search	4.1.5	102
e_f	equal-frequency discretization	4.2.3	132
e_w	equal-width discretization	4.2.2	130
exchange	generate neighbor clustering	4.1.5	102
GENSAM	generic sampling	4.3.1	152
IBL2	instance-based learning	4.1.6	105
initialize	initialize clustering / similarity threshold δ	4.1.5, 4.2.3	102, 135
LEACLU	leader clustering	4.1.4	100
LEASAM	leader sampling	4.2.2	116
LEASAM$^+$	modified leader sampling	4.2.3	135
NN	nearest neighbor classifier	2.5.1	37
partition	partition tuples	4.1.7	108
PL	prototyping algorithm	4.1.7	108

Algorithm	Description	Reference	Page
prototype	generate prototype	4.1.7	108
quality	compute clustering quality	4.1.5	102
random	generate random number	4.1.1	88
RANSAM	simple random sampling	4.1.1	88
reduce	remove less representative prototypes	4.2.5	147
relevance	compute attribute relevance weights	4.3.1	152
SHUFFLE	shuffle focusing input into random order	4.3.1	152
SIMSAM	similarity-driven sampling	4.2.3	135
SORT	sort focusing input	4.3.1	152
STRATIFY	stratification	4.2.1	123
STRSAM	stratified sampling	4.1.3	90
SYSSAM	systematic sampling	4.1.2	89
update	update similarity threshold δ	4.2.4	144

Appendix B

More Evaluation Criteria

In this appendix, we specify definitions of remaining filter and wrapper evaluation criteria in isolated and comparative modes as well as cumulated evaluation criteria in comparative mode. For descriptions and explanations, we refer to section 3.4.3.

B.1 Filter Evaluation Criteria

In this section, we specify remaining filter evaluation criteria.

Definition B.1.1 (Evaluation Criterion $E_{V_j}^{\leftrightarrow}$)

Assume a (database) table (A, T), a quantitative attribute $a_j \in A$, a focusing input $f_{in} \subseteq T$, a focusing solution $F : T^{\subseteq} \longrightarrow T^{\subseteq}$, a focusing output $f_{out} = F(f_{in}) \subseteq f_{in}$, if F is deterministic, or a set of focusing outputs F_{out}, $\forall f_{out} \in F_{out} : f_{out} = F(f_{in}) \subseteq f_{in}$, if F is non-deterministic, an additional value $\sigma(q_{V_j}) \geq 0$ that depicts the fraction of the sum of s_{V_j} and its standard deviation and d_{V_j} in case of non-deterministic focusing solutions and their averaged application, $0 < \alpha < 1$, and $0 \leq \eta_\sigma \leq 0.5$. We define the evaluation criterion $E_{V_j}^{\leftrightarrow}$ as a function

$$E_{V_j}^{\leftrightarrow} : F \times T^{\subseteq} \times T^{\subseteq} \longrightarrow [0; \infty[.$$

If

$$
\begin{array}{llll}
(i) & :\Longleftrightarrow & F \text{ deterministic} \quad \text{and} & \mid q_{V_j}(\alpha, f_{in}, f_{out}) \mid \leq 1, \\
(ii) & :\Longleftrightarrow & F \text{ non-deterministic} \quad \text{and} & \mid q_{V_j}(\alpha, f_{in}, f_{out}) \mid \leq 1 \\
& & & \text{in most cases,}
\end{array}
$$

T. Reinartz: Focusing Solutions for Data Mining, LNAI 1623, pp. 267-276, 1999
© Springer-Verlag Berlin Heidelberg 1999

(iii) $:\Longleftrightarrow$ F deterministic   and $|q_{V_j}(\alpha, f_{in}, f_{out})| > 1,$

(iv) $:\Longleftrightarrow$ F non-deterministic and $|q_{V_j}(\alpha, f_{in}, f_{out})| > 1$

in most cases,

and

$E_{V_j}(F, f_{in}, f_{out})$

$$
:= \begin{cases}
0, & if\ (i) \\[2mm]
1 - \dfrac{|\{f_{out} \in F_{out}\ |\ |q_{V_j}(\alpha, f_{in}, f_{out})| \le 1)\}|}{|\{F_{out}\}|}, & if\ (ii) \\[3mm]
1, & if\ (iii) \\[2mm]
0 + \dfrac{|\{f_{out} \in F_{out}\ |\ |q_{V_j}(\alpha, f_{in}, f_{out})| > 1)\}|}{|\{F_{out}\}|}, & if\ (iv)
\end{cases}
$$

$E_{V_j}^{\leftrightarrow}(F, f_{in}, f_{out})$

$$
\begin{aligned}
:= \quad & \tfrac{1}{2} \cdot E_{V_j}(F, f_{in}, f_{out}) \\
+ \quad & \tfrac{1}{2} \cdot \Big((1 - \eta_\sigma) \cdot |q_{V_j}(\alpha, f_{in}, f_{out})| \\
& \qquad + \eta_\sigma \cdot |\sigma(q_{V_j}(\alpha, f_{in}, f_{out}))| \Big).
\end{aligned}
$$

Definition B.1.2 (Evaluation Criterion $E_{D_j}^{\leftrightarrow}$)

Assume a (database) table (A, T), an attribute $a_j \in A$, a focusing input $f_{in} \subseteq T$, a focusing solution $F : T^{\subseteq} \longrightarrow T^{\subseteq}$, a focusing output $f_{out} = F(f_{in}) \subseteq f_{in}$, if F is deterministic, or a set of focusing outputs F_{out}, $\forall f_{out} \in F_{out} : f_{out} = F(f_{in}) \subseteq f_{in}$, if F is non-deterministic, an additional value $\sigma(q_{D_j}) \ge 0$ that depicts the fraction of the sum of s_{D_j} and its standard deviation and d_{D_j} in case of non-deterministic focusing solutions and their averaged application, $0 < \alpha < 1$, and $0 \le \eta_\sigma \le 0.5$. We define the evaluation criterion $E_{D_j}^{\leftrightarrow}$ as a function

$$
E_{D_j}^{\leftrightarrow} : F \times T^{\subseteq} \times T^{\subseteq} \longrightarrow [0;\, \infty[.
$$

If

(i) $:\Longleftrightarrow$ F deterministic   and $|q_{D_j}(\alpha, f_{in}, f_{out})| \le 1,$

(ii) $:\Longleftrightarrow$ F non-deterministic and $|q_{D_j}(\alpha, f_{in}, f_{out})| \le 1$

in most cases,

(iii) $:\iff$ F deterministic   and $|\,q_{D_j}(\alpha, f_{in}, f_{out})\,|\, > 1,$

(iv) $:\iff$ F non-deterministic   and $|\,q_{D_j}(\alpha, f_{in}, f_{out})\,|\, > 1$

in most cases,

and

$E_{D_j}(F, f_{in}, f_{out})$

$$
:= \begin{cases}
0, & \text{if (i)} \\[2mm]
1 - \dfrac{|\,\{f_{out} \in F_{out} \mid |\, q_{D_j}(\alpha, f_{in}, f_{out})\,|\, \le 1)\}\,|}{|\,\{F_{out}\}\,|}, & \text{if (ii)} \\[4mm]
1, & \text{if (iii)} \\[2mm]
0 + \dfrac{|\,\{f_{out} \in F_{out} \mid |\, q_{D_j}(\alpha, f_{in}, f_{out})\,|\, > 1)\}\,|}{|\,\{F_{out}\}\,|}, & \text{if (iv)}
\end{cases}
$$

$E_{D_j}^{\leftrightarrow}(F, f_{in}, f_{out})$

$:=$ $\tfrac{1}{2} \cdot E_{D_j}(F, f_{in}, f_{out})$

$+$ $\tfrac{1}{2} \cdot \Big((1 - \eta_\sigma) \cdot |\, q_{D_j}(\alpha, f_{in}, f_{out})\,|$

$+$ $\eta_\sigma \cdot |\, \sigma(q_{D_j}(\alpha, f_{in}, f_{out}))\,|\,\Big).$

Definition B.1.3 (Evaluation Criterion $E_J^{\leftrightarrow}$)

Assume a (database) table (A, T), an attribute $a_j \in A$, a focusing input $f_{in} \subseteq T$, a focusing solution $F : T^{\subseteq} \longrightarrow T^{\subseteq}$, a focusing output $f_{out} = F(f_{in}) \subseteq f_{in}$, if F is deterministic, or a set of focusing outputs F_{out}, $\forall f_{out} \in F_{out} : f_{out} = F(f_{in}) \subseteq f_{in}$, if F is non-deterministic, an additional value $\sigma(q_J) \ge 0$ that depicts the fraction of the sum of s_J and its standard deviation and d_J in case of non-deterministic focusing solutions and their averaged application, $0 < \alpha < 1$, and $0 \le \eta_\sigma \le 0.5$. We define the evaluation criterion $E_J^{\leftrightarrow}$ as a function

$E_J^{\leftrightarrow} : F \times T^{\subseteq} \times T^{\subseteq} \longrightarrow [0; \infty[.$

If

(i) $:\iff$ F deterministic   and $|\,q_J(\alpha, f_{in}, f_{out})\,|\, \le 1,$

(ii) $:\iff$ F non-deterministic   and $|\,q_J(\alpha, f_{in}, f_{out})\,|\, \le 1$

in most cases,

$$(iii) \quad :\Longleftrightarrow \quad F \text{ deterministic} \qquad \text{and} \quad \mid q_J(\alpha, f_{in}, f_{out}) \mid > 1,$$

$$(iv) \quad :\Longleftrightarrow \quad F \text{ non-deterministic} \quad \text{and} \quad \mid q_J(\alpha, f_{in}, f_{out}) \mid > 1$$

$$\text{in most cases,}$$

and

$$E_J(F, f_{in}, f_{out})$$

$$:= \begin{cases} 0, & \text{if } (i) \\[2mm] 1 - \dfrac{\mid \{f_{out} \in F_{out} \mid \mid q_J(\alpha, f_{in}, f_{out}) \mid \leq 1)\} \mid}{\mid \{F_{out}\} \mid}, & \text{if } (ii) \\[2mm] 1, & \text{if } (iii) \\[2mm] 0 + \dfrac{\mid \{f_{out} \in F_{out} \mid \mid q_J(\alpha, f_{in}, f_{out}) \mid > 1)\} \mid}{\mid \{F_{out}\} \mid}, & \text{if } (iv) \end{cases}$$

$$E_J^{\leftrightarrow}(F, f_{in}, f_{out})$$

$$:= \quad \tfrac{1}{2} \cdot E_J(F, f_{in}, f_{out})$$

$$+ \quad \tfrac{1}{2} \cdot \left((1 - \eta_\sigma) \cdot \mid q_J(\alpha, f_{in}, f_{out}) \mid \right.$$

$$\left. + \ \eta_\sigma \cdot \mid \sigma(q_J(\alpha, f_{in}, f_{out})) \mid \right).$$

Definition B.1.4 (Evaluation Criterion $E_{V_j}^{\updownarrow}$)

Assume a (database) table (A, T), a quantitative attribute $a_j \in A$, a focusing input $f_{in} \subseteq T$, a focusing solution $F : T^{\subseteq} \longrightarrow T^{\subseteq}$, a focusing output $f_{out} = F(f_{in}) \subseteq f_{in}$, if F is deterministic, or a set of focusing outputs F_{out}, $\forall f_{out} \in F_{out} : f_{out} = F(f_{in}) \subseteq f_{in}$, if F is non-deterministic, a set of focusing outputs F'_{out}, $\forall f'_{out} \in F'_{out} : f'_{out} \subseteq f_{in}$, additional values $\sigma(q_{V_j}) \geq 0$ that depict the fraction of the sum of s_{V_j} and its standard deviation and d_{V_j} in case of non-deterministic focusing solutions and their averaged application, $0 < \alpha < 1$, and $0 \leq \eta_\sigma \leq 0.5$. We define the evaluation criterion $E_{V_j}^{\updownarrow}$ as a function

$$E_{V_j}^{\updownarrow} : F \times T^{\subseteq} \times T^{\subseteq} \times \{T^{\subseteq}, \dots, T^{\subseteq}\} \longrightarrow [0; \infty[,$$

$$E_{V_j}^{\uparrow}(F, f_{in}, f_{out}, F'_{out})$$

$$:= \quad \tfrac{1}{2} \cdot E_{V_j}(F, f_{in}, f_{out})$$

$$+ \quad \tfrac{1}{2} \cdot \left((1 - \eta_\sigma) \cdot \frac{\mid q_{V_j}(\alpha, f_{in}, f_{out}) \mid}{\displaystyle\max_{f'_{out} \in F'_{out}} \{\mid q_{V_j}(\alpha, f_{in}, f'_{out}) \mid\}} \right.$$

$$\left. + \quad \eta_\sigma \cdot \frac{\mid \sigma(q_{V_j}(\alpha, f_{in}, f_{out})) \mid}{\displaystyle\max_{f'_{out} \in F'_{out}} \{\mid \sigma(q_{V_j}(\alpha, f_{in}, f'_{out})) \mid\}} \right).$$

Definition B.1.5 (Evaluation Criterion $E_{D_j}^{\uparrow}$)

Assume a (database) table (A, T), an attribute $a_j \in A$, a focusing input $f_{in} \subseteq T$, a focusing solution $F : T^{\subseteq} \longrightarrow T^{\subseteq}$, a focusing output $f_{out} = F(f_{in}) \subseteq f_{in}$, if F is deterministic, or a set of focusing outputs F_{out}, $\forall f_{out} \in F_{out} : f_{out} = F(f_{in}) \subseteq f_{in}$, if F is non-deterministic, a set of focusing outputs F'_{out}, $\forall f'_{out} \in F'_{out} : f'_{out} \subseteq f_{in}$, additional values $\sigma(q_{D_j}) \geq 0$ that depict the fraction of the sum of s_{D_j} and its standard deviation and d_{D_j} in case of non-deterministic focusing solutions and their averaged application, $0 < \alpha < 1$, and $0 \leq \eta_\sigma \leq 0.5$. We define the evaluation criterion $E_{D_j}^{\uparrow}$ as a function

$$E_{D_j}^{\uparrow} : F \times T^{\subseteq} \times T^{\subseteq} \times \{T^{\subseteq}, \ldots, T^{\subseteq}\} \longrightarrow [0; \infty[,$$

$$E_{D_j}^{\uparrow}(F, f_{in}, f_{out}, F'_{out})$$

$$:= \quad \tfrac{1}{2} \cdot E_{D_j}(F, f_{in}, f_{out})$$

$$+ \quad \tfrac{1}{2} \cdot \left((1 - \eta_\sigma) \cdot \frac{\mid q_{D_j}(\alpha, f_{in}, f_{out}) \mid}{\displaystyle\max_{f'_{out} \in F'_{out}} \{\mid q_{D_j}(\alpha, f_{in}, f'_{out}) \mid\}} \right.$$

$$\left. + \quad \eta_\sigma \cdot \frac{\mid \sigma(q_{D_j}(\alpha, f_{in}, f_{out})) \mid}{\displaystyle\max_{f'_{out} \in F'_{out}} \{\mid \sigma(q_{D_j}(\alpha, f_{in}, f'_{out})) \mid\}} \right).$$

Definition B.1.6 (Evaluation Criterion $E_J^{\uparrow}$)

Assume a (database) table (A, T), an attribute $a_j \in A$, a focusing input $f_{in} \subseteq T$, a focusing solution $F : T^{\subseteq} \longrightarrow T^{\subseteq}$, a focusing output $f_{out} = F(f_{in}) \subseteq f_{in}$, if F is deterministic, or a set of focusing outputs F_{out}, $\forall f_{out} \in F_{out} : f_{out} = F(f_{in}) \subseteq f_{in}$, if F is non-deterministic, a set of focusing outputs F'_{out}, $\forall f'_{out} \in F'_{out} : f'_{out} \subseteq f_{in}$, additional values $\sigma(q_J) \geq 0$ that depict the fraction of the sum of s_J and its standard deviation and d_J in case of non-deterministic focusing solutions and their averaged application, $0 < \alpha < 1$, and $0 \leq \eta_\sigma \leq 0.5$. We define the evaluation criterion $E_J^{\uparrow}$ as a function

$$E_J^\updownarrow : F \times T^\subseteq \times T^\subseteq \times \{T^\subseteq, \ldots, T^\subseteq\} \longrightarrow [0; \infty[,$$

$$E_J^\updownarrow(F, f_{in}, f_{out}, F'_{out})$$

$$:= \quad \tfrac{1}{2} \cdot E_J(F, f_{in}, f_{out})$$

$$+ \quad \tfrac{1}{2} \cdot \left((1 - \eta_\sigma) \cdot \frac{\mid q_J(\alpha, f_{in}, f_{out}) \mid}{\displaystyle\max_{f'_{out} \in F'_{out}} \{\mid q_J(\alpha, f_{in}, f'_{out}) \mid\}} \right.$$

$$\left. + \quad \eta_\sigma \cdot \frac{\mid \sigma(q_J(\alpha, f_{in}, f_{out})) \mid}{\displaystyle\max_{f'_{out} \in F'_{out}} \{\mid \sigma(q_J(\alpha, f_{in}, f'_{out})) \mid\}} \right).$$

Definition B.1.7 (Evaluation Criterion $E_F^\updownarrow$)

Assume a (database) table (A, T), a focusing input $f_{in} \subseteq T$, a focusing solution $F : T^\subseteq \longrightarrow T^\subseteq$, a focusing output $f_{out} = F(f_{in}) \subseteq f_{in}$, if F is deterministic, or a set of focusing outputs F_{out}, $\forall f_{out} \in F_{out} : f_{out} = F(f_{in}) \subseteq f_{in}$, if F is non-deterministic, a set of focusing outputs F'_{out}, $\forall f'_{out} \in F'_{out} : f'_{out} \subseteq f_{in}$, and $0 \leq \eta_S, \eta_V, \eta_D, \eta_J \leq 1$ with $\eta_S + \eta_V + \eta_D + \eta_J = 1$. We define the evaluation criterion $E_F^\updownarrow$ as a function

$$E_F^\updownarrow : F \times T^\subseteq \times T^\subseteq \times \{T^\subseteq, \ldots, T^\subseteq\} \longrightarrow [0; \infty[,$$

$$E_F^\updownarrow(F, f_{in}, f_{out}, F'_{out})$$

$$:= \quad \eta_S \cdot \sum_{j=1}^{N} E_{S_j}^\updownarrow(F, f_{in}, f_{out}, F'_{out}) \; + \; \eta_V \cdot \sum_{j=1}^{N} E_{V_j}^\updownarrow(F, f_{in}, f_{out}, F'_{out})$$

$$+ \quad \eta_D \cdot \sum_{j=1}^{N} E_{D_j}^\updownarrow(F, f_{in}, f_{out}, F'_{out}) \; + \; \eta_J \cdot E_J^\updownarrow(F, f_{in}, f_{out}, F'_{out}).$$

B.2   Wrapper Evaluation Criteria

In this section, we specify remaining wrapper evaluation criteria.

Definition B.2.1 (Evaluation Criterion $E_M^\leftrightarrow$)

Assume a (database) table (A, T), a focusing input $f_{in} \subseteq T$, a focusing solution $F : T^\subseteq \longrightarrow T^\subseteq$, a focusing output $f_{out} = F(f_{in}) \subseteq f_{in}$, if F is deterministic, or a set of focusing outputs F_{out}, $\forall f_{out} \in F_{out} : f_{out} = F(f_{in}) \subseteq f_{in}$, if F is non-deterministic, a data mining algorithm D, an additional value $\sigma(m) \geq 0$ that depicts the sum of m and its standard deviation in case of non-deterministic *focusing solutions and their averaged application, and $0 \leq \eta_\sigma \leq 0.5$. We define the evaluation criterion $E_M^{\leftrightarrow}$ as a function*

$$E_M^{\leftrightarrow} : D \times T^{\subseteq} \times T^{\subseteq} \longrightarrow]0; \infty[,$$

$$E_M^{\leftrightarrow}(D, f_{in}, f_{out})$$

$$:= (1 - \eta_\sigma) \cdot \frac{m(D, f_{out})}{m(D, f_{in})} + \eta_\sigma \cdot \frac{\sigma\left(m(D, f_{out})\right)}{m(D, f_{in})}.$$

Definition B.2.2 (Evaluation Criterion $E_A^{\leftrightarrow}$)

Assume a (database) table (A, T), a focusing input $f_{in} \subseteq T$, a focusing solution $F : T^{\subseteq} \longrightarrow T^{\subseteq}$, a focusing output $f_{out} = F(f_{in}) \subseteq f_{in}$, if F is deterministic, or a set of focusing outputs F_{out}, $\forall f_{out} \in F_{out} : f_{out} = F(f_{in}) \subseteq f_{in}$, if F is non-deterministic, a data mining algorithm D, an additional value $\sigma(e) \geq 0$ that depicts the sum of e and its standard deviation in case of non-deterministic focusing solutions and their averaged application, a test set $T_{test} \subseteq T$, and $0 \leq \eta_\sigma \leq 0.5$. We define the evaluation criterion $E_A^{\leftrightarrow}$ as a function

$$E_A^{\leftrightarrow} : D \times T^{\subseteq} \times T^{\subseteq} \times T^{\subseteq} \longrightarrow]0; \infty[,$$

$$E_A^{\leftrightarrow}(D, f_{in}, f_{out}, T_{test})$$

$$:= (1 - \eta_\sigma) \cdot \left(\frac{e\left(D(f_{out}), T_{test}\right)}{e\left(D(f_{in}), T_{test}\right)}\right) + \eta_\sigma \cdot \left(\frac{\sigma\left(e\left(D(f_{out}), T_{test}\right)\right)}{e\left(D(f_{in}), T_{test}\right)}\right).$$

Definition B.2.3 (Evaluation Criterion $E_C^{\leftrightarrow}$)

Assume a (database) table (A, T), a focusing input $f_{in} \subseteq T$, a focusing solution $F : T^{\subseteq} \longrightarrow T^{\subseteq}$, a focusing output $f_{out} = F(f_{in}) \subseteq f_{in}$, if F is deterministic, or a set of focusing outputs F_{out}, $\forall f_{out} \in F_{out} : f_{out} = F(f_{in}) \subseteq f_{in}$, if F is non-deterministic, a data mining algorithm D, and $0 \leq \eta_\sigma \leq 0.5$. We define the evaluation criterion $E_C^{\leftrightarrow}$ as a function

$$E_C^{\leftrightarrow} : D \times T^{\subseteq} \times T^{\subseteq} \longrightarrow]0; \infty[.$$

If D is top down induction of decision trees, assume two additional values $\sigma(l) \geq 0$ and $\sigma(n) \geq 0$ that depict the sum of l and its standard deviation and the sum of n and its standard deviation, respectively, in case of non-deterministic focusing solutions and their averaged application,

$$E_C^{\leftrightarrow}(D, f_{in}, f_{out})$$

$$:= \frac{1}{2} \cdot \left((1 - \eta_\sigma) \cdot \frac{l\,(D(f_{out}))}{l\,(D(f_{in}))} + \eta_\sigma \cdot \frac{\sigma\,(l\,(D(f_{out})))}{l\,(D(f_{in}))} \right)$$

$$+ \frac{1}{2} \cdot \left((1 - \eta_\sigma) \cdot \frac{n\,(D(f_{out}))}{n\,(D(f_{in}))} + \eta_\sigma \cdot \frac{\sigma\,(n\,(D(f_{out})))}{n\,(D(f_{in}))} \right).$$

If D is nearest neighbor,

$$E_C^{\leftrightarrow}(D, f_{in}, f_{out}) \quad := \quad E_M^{\leftrightarrow}(D, f_{in}, f_{out}).$$

Definition B.2.4 (Evaluation Criterion $E_M^{\updownarrow}$)

Assume a (database) table (A, T), a focusing input $f_{in} \subseteq T$, a focusing solution $F : T^{\subseteq} \longrightarrow T^{\subseteq}$, a focusing output $f_{out} = F(f_{in}) \subseteq f_{in}$, if F is deterministic, or a set of focusing outputs F_{out}, $\forall f_{out} \in F_{out} : f_{out} = F(f_{in}) \subseteq f_{in}$, if F is non-deterministic, a set of focusing outputs F'_{out}, $\forall f'_{out} \in F'_{out} : f'_{out} \subseteq f_{in}$, a data mining algorithm D, an additional value $\sigma(m) \geq 0$ that depicts the sum of m and its standard deviation in case of non-deterministic focusing solutions and their averaged application, and $0 \leq \eta_\sigma \leq 0.5$. We define the evaluation criterion $E_M^{\updownarrow}$ as a function

$$E_M^{\updownarrow} : D \times T^{\subseteq} \times \{T^{\subseteq}, \dots, T^{\subseteq}\} \longrightarrow \,]0;\,1],$$

$$E_M^{\updownarrow}(D, f_{out}, F'_{out})$$

$$:= (1 - \eta_\sigma) \cdot \frac{m(D, f_{out})}{\max\limits_{f'_{out} \in F'_{out}} \{m(D, f'_{out})\}} + \eta_\sigma \cdot \frac{\sigma\,(m(D, f_{out}))}{\max\limits_{f'_{out} \in F'_{out}} \{\sigma\,(m(D, f'_{out}))\}}.$$

Definition B.2.5 (Evaluation Criterion $E_A^{\updownarrow}$)

Assume a (database) table (A, T), a focusing input $f_{in} \subseteq T$, a focusing solution $F : T^{\subseteq} \longrightarrow T^{\subseteq}$, a focusing output $f_{out} = F(f_{in}) \subseteq f_{in}$, if F is deterministic, or a set of focusing outputs F_{out}, $\forall f_{out} \in F_{out} : f_{out} = F(f_{in}) \subseteq f_{in}$, if F is non-deterministic, a set of focusing outputs F'_{out}, $\forall f'_{out} \in F'_{out} : f'_{out} \subseteq f_{in}$, a data mining algorithm D, an additional value $\sigma(e) \geq 0$ that depicts the sum of e and its standard deviation in case of non-deterministic focusing solutions and their averaged application, a test set $T_{test} \subseteq T$, and $0 \leq \eta_\sigma \leq 0.5$. We define the evaluation criterion $E_A^{\updownarrow}$ as a function

$$E_A^{\updownarrow} : D \times T^{\subseteq} \times \{T^{\subseteq}, \dots, T^{\subseteq}\} \times T^{\subseteq} \longrightarrow \,]0;\,1],$$

$$E_A^\updownarrow(D, f_{out}, F'_{out}, T_{test})$$

$$:= (1 - \eta_\sigma) \cdot \frac{e(D(f_{out}), T_{test})}{\max_{f'_{out} \in F'_{out}} \{e(D(f'_{out}), T_{test}))\}}$$

$$+ \eta_\sigma \cdot \frac{\sigma(e(D(f_{out}), T_{test})))}{\max_{f'_{out} \in F'_{out}} \{\sigma(e(D(f'_{out}), T_{test})))\}}.$$

Definition B.2.6 (Evaluation Criterion $E_C^\updownarrow$)

Assume a (database) table (A, T), a focusing input $f_{in} \subseteq T$, a focusing solution $F : T^\subseteq \longrightarrow T^\subseteq$, a focusing output $f_{out} = F(f_{in}) \subseteq f_{in}$, if F is deterministic, or a set of focusing outputs F_{out}, $\forall f_{out} \in F_{out} : f_{out} = F(f_{in}) \subseteq f_{in}$, if F is non-deterministic, a set of focusing outputs F'_{out}, $\forall f'_{out} \in F'_{out} : f'_{out} \subseteq f_{in}$, a data mining algorithm D, and $0 \le \eta_\sigma \le 0.5$. We define the evaluation criterion $E_C^\updownarrow$ as a function

$$E_C^\updownarrow : D \times T^\subseteq \times \{T^\subseteq, \dots, T^\subseteq\} \longrightarrow]0; 1].$$

If D is top down induction of decision trees, assume two additional values $\sigma(l) \ge 0$ and $\sigma(n) \ge 0$ that depict the sum of l and its standard deviation and the sum of n and its standard deviation, respectively, in case of non-deterministic focusing solutions and their averaged application,

$$E_C^\updownarrow(D, f_{out}, F'_{out})$$

$$:= \frac{1}{2} \cdot \left((1 - \eta_\sigma) \cdot \frac{l(D(f_{out}))}{\max_{f'_{out} \in F'_{out}} \{l(D(f'_{out}))\}} \right.$$

$$+ \eta_\sigma \cdot \frac{\sigma(l(D(f_{out})))}{\max_{f'_{out} \in F'_{out}} \{\sigma(l(D(f'_{out})))\}} \right)$$

$$+ \frac{1}{2} \cdot \left((1 - \eta_\sigma) \cdot \frac{n(D(f_{out}))}{\max_{f'_{out} \in F'_{out}} \{n(D(f'_{out}))\}} \right.$$

$$+ \eta_\sigma \cdot \frac{\sigma(n(D(f_{out})))}{\max_{f'_{out} \in F'_{out}} \{\sigma(n(D(f'_{out})))\}} \right).$$

If D is nearest neighbor,

$$E_C^\updownarrow(D, f_{in}, F'_{out}) \quad := \quad E_M^\updownarrow(D, f_{in}, F'_{out}).$$

Definition B.2.7 (Evaluation Criterion $E_W^{\uparrow}$)

Assume a (database) table (A, T), a focusing input $f_{in} \subseteq T$, a focusing solution $F : T^{\subseteq} \longrightarrow T^{\subseteq}$, a focusing output $f_{out} = F(f_{in}) \subseteq f_{in}$, if F is deterministic, or a set of focusing outputs F_{out}, $\forall f_{out} \in F_{out} : f_{out} = F(f_{in}) \subseteq f_{in}$, if F is non-deterministic, a set of focusing outputs F_{out}', $\forall f_{out}' \in F_{out}' : f_{out}' \subseteq f_{in}$, a data mining algorithm D, a test set $T_{test} \subseteq T$, and $0 \le \eta_T, \eta_M, \eta_A, \eta_C \le 1$ with $\eta_T + \eta_M + \eta_A + \eta_C = 1$. We define the evaluation criterion $E_W^{\uparrow}$ as a function

$$E_W^{\uparrow} : F \times D \times T^{\subseteq} \times T^{\subseteq} \times \{T^{\subseteq}, \dots, T^{\subseteq}\} \times T^{\subseteq} \longrightarrow \;]0; 1],$$

$$E_W^{\uparrow}(F, D, f_{in}, f_{out}, F_{out}', T_{test})$$

$$\begin{aligned} := \quad & \eta_T \cdot E_T^{\uparrow}(F, D, f_{in}, f_{out}, F_{out}') \; + \; \eta_M \cdot E_M^{\uparrow}(D, f_{out}, F_{out}') \\ + \quad & \eta_A \cdot E_A^{\uparrow}(D, f_{out}, F_{out}', T_{test}) \; + \; \eta_C \cdot E_C^{\uparrow}(D, f_{out}, F_{out}'). \end{aligned}$$

Appendix C

Remaining Proofs

In this appendix, we show remaining proofs for propositions 2.5.2 (page 41) and 4.2.3 (page 119). The first proposition states that the distance measure which corresponds to the weighted cumulated similarity measure meets the triangle inequality, and the second proposition establishes similarity in sorted tables.

Proposition C.0.1 (Δ-Inequality)

Assume a (database) table (A, T), and a weighted cumulated similarity measure Sim. Then,

$$\forall t_i, t_{i'}, t_{i''} \in T : \quad 1 - Sim(t_i, t_{i'}) + 1 - Sim(t_{i'}, t_{i''}) \geq 1 - Sim(t_i, t_{i''})$$

Proof:

First, we show

$$\forall a_j \in A \; \forall t_i, t_{i'}, t_{i''} \in T :$$

$$sim_j(t_{ij}, t_{i'j}) + sim_j(t_{i'j}, t_{i''j}) - sim_j(t_{ij}, t_{i''j}) \leq 1,$$

and then we use this property to prove the Δ-inequality.

Assume a_j is qualitative, and $sim_j(t_{ij}, t_{i'j}) + sim_j(t_{i'j}, t_{i''j}) - sim_j(t_{ij}, t_{i''j}) > 1$. Since sim_j distinguishes only two cases and local similarity between two qualitative values is either 0 or 1, the only situation with $sim_j(t_{ij}, t_{i'j}) + sim_j(t_{i'j}, t_{i''j}) - sim_j(t_{ij}, t_{i''j}) > 1$ is $sim_j(t_{ij}, t_{i'j}) = 1$, $sim_j(t_{i'j}, t_{i''j}) = 1$, and $sim_j(t_{ij}, t_{i''j}) = 0$. This means by proposition 2.5.1 $t_{ij} = t_{i'j}$, $t_{i'j} = t_{i''j}$, and $t_{ij} \neq t_{i''j}$. On the other hand, the first two equalities imply $t_{ij} = t_{i''j}$ in contradiction to $t_{ij} \neq t_{i''j}$.

Assume a_j is quantitative, and a distance measure

$$d : dom(a_j) \times dom(a_j) \longrightarrow [0; 1], \quad d(a_{jk}, a_{jk'}) := \frac{\mid a_{jk} - a_{jk'} \mid}{r_j(T)}.$$

T. Reinartz: Focusing Solutions for Data Mining, LNAI 1623, pp. 277-280, 1999
© Springer-Verlag Berlin Heidelberg 1999

Since d is a metric, we have

$$\forall a_{jk}, a_{jk'}, a_{jk''} \in dom(a_j) : d(a_{jk}, a_{jk'}) + d(a_{jk'}, a_{jk''}) \geq d(a_{jk}, a_{jk''}).$$

Hence, we also have

$$\forall a_j \in A \ \forall t_i, t_{i'}, t_{i''} \in T :$$

$$sim_j(t_{ij}, t_{i'j}) + sim_j(t_{i'j}, t_{i''j}) - sim_j(t_{ij}, t_{i''j}) \leq 1$$

for quantitative attributes.

If we use this property and apply definition 2.5.9, we infer the Δ-inequality in the following way:

$$1 = \frac{1}{w} \cdot \sum_{j=1}^{N} w_j$$

$$\geq \frac{1}{w} \cdot \sum_{j=1}^{N} w_j \cdot (sim_j(t_{ij}, t_{i'j}) + sim_j(t_{i'j}, t_{i''j}) - sim_j(t_{ij}, t_{i''j}))$$

$$\Longleftrightarrow \quad 1 \geq \frac{1}{w} \cdot \sum_{j=1}^{N} w_j \cdot sim_j(t_{ij}, t_{i'j}) + \frac{1}{w} \cdot \sum_{j=1}^{N} w_j \cdot sim_j(t_{i'j}, t_{i''j})$$

$$- \frac{1}{w} \cdot \sum_{j=1}^{N} w_j \cdot sim_j(t_{ij}, t_{i''j})$$

$$\Longleftrightarrow \quad 1 \geq Sim(t_i, t_{i'}) + Sim(t_{i'}, t_{i''}) - Sim(t_i, t_{i''})$$

$$\Longleftrightarrow \quad 1 - Sim(t_i, t_{i'}) + 1 - Sim(t_{i'}, t_{i''}) \geq 1 - Sim(t_i, t_{i''})$$

$$\square$$

Proposition C.0.2 (Similarity in Sorted Tables)

Assume a (database) table (A, T).

$$\forall t_i, t_{i'}, t_{i''} \in T :$$

$$\succ (t_i, t_{i'}) \wedge \ \succ (t_{i'}, t_{i''}) \wedge \sum_{j=1}^{j^*_{ii'}} w_j \cdot sim_j(t_{ij}, t_{i'j}) \geq \sum_{j=1}^{j^*_{ii'}} w_j \cdot sim_j(t_{ij}, t_{i''j})$$

$$\Longrightarrow \quad Sim(t_i, t_{i'}) > Sim(t_i, t_{ii''})$$

Proof:

$$Sim(t_i, t_{i'})$$

$$\overset{Def.\ 2.5.9}{=} \quad \frac{1}{w} \cdot \sum_{j=1}^{N} w_j \cdot sim_j(t_{ij}, t_{i'j})$$

$\underset{=}{Prop.\ 4.2.2\ (i)}$

$$\left(\frac{1}{w} \cdot \sum_{j=1}^{j^*_{ii'}-1} w_j \cdot sim_j(t_{ij}, t_{i'j}) \right)$$

$$+ \quad \frac{1}{w} \cdot w_{j^*_{ii'}} \cdot sim_{j^*_{ii'}}(t_{ij^*_{ii'}}, t_{i'j^*_{ii'}})$$

$$+ \quad \left(\frac{1}{w} \cdot \sum_{j=j^*_{ii'}+1}^{j^*_{ii''}-1} w_j \cdot sim_j(t_{ij}, t_{i'j}) \right)$$

$$+ \quad \frac{1}{w} \cdot w_{j^*_{ii''}} \cdot sim_{j^*_{ii''}}(t_{ij^*_{ii''}}, t_{i'j^*_{ii''}})$$

$$+ \quad \left(\frac{1}{w} \cdot \sum_{j=j^*_{ii''}+1}^{N} w_j \cdot sim_j(t_{ij}, t_{i'j}) \right)$$

$\underset{=}{Def.\ 4.2.2}$

$$\left(\frac{1}{w} \cdot \sum_{j=1}^{j^*_{ii'}-1} w_j \cdot sim_j(t_{ij}, t_{i'j}) \right)$$

$$+ \quad \frac{1}{w} \cdot w_{j^*_{ii'}} \cdot sim_{j^*_{ii'}}(t_{ij^*_{ii'}}, t_{i'j^*_{ii'}})$$

$$+ \quad \left(\frac{1}{w} \cdot \sum_{j=j^*_{ii'}+1}^{j^*_{ii''}-1} w_j \right)$$

$$+ \quad \frac{1}{w} \cdot w_{j^*_{ii''}}$$

$$+ \quad \left(\frac{1}{w} \cdot \sum_{j=j^*_{ii''}+1}^{N} w_j \right)$$

$\underset{\geq}{Precondition}$

$$\left(\frac{1}{w} \cdot \sum_{j=1}^{j^*_{ii'}-1} w_j \cdot sim_j(t_{ij}, t_{i''j}) \right)$$

$$+ \quad \frac{1}{w} \cdot w_{j^*_{ii'}} \cdot sim_{j^*_{ii'}}(t_{ij^*_{ii'}}, t_{i''j^*_{ii'}})$$

$$+ \quad \left(\frac{1}{w} \cdot \sum_{j=j^*_{ii'}+1}^{j^*_{ii''}-1} w_j \right)$$

$$+ \quad \frac{1}{w} \cdot w_{j^*_{ii''}}$$

$$+ \quad \left(\frac{1}{w} \cdot \sum_{j=j^*_{ii''}+1}^{N} w_j \right)$$

$\underset{>}{Def.\ 4.2.2}$

$$\left(\frac{1}{w} \cdot \sum_{j=1}^{j^*_{ii'}-1} w_j \cdot sim_j(t_{ij}, t_{i''j}) \right)$$

$$+ \quad \frac{1}{w} \cdot w_{j^*_{ii'}} \cdot sim_{j^*_{ii'}}(t_{ij^*_{ii'}}, t_{i''j^*_{ii'}})$$

$$+ \quad \left(\frac{1}{w} \cdot \sum_{j=j^*_{ii'}+1}^{j^*_{ii''}-1} w_j \right)$$

$$+ \quad \frac{1}{w} \cdot w_{j^*_{ii''}} \cdot sim_{j^*_{ii''}}(t_{ij^*_{ii''}}, t_{i''j^*_{ii''}})$$

$$+ \quad \left(\frac{1}{w} \cdot \sum_{j=j^*_{ii''}+1}^{N} w_j \right)$$

$$\overset{Def.\ 2.5.7\ and\ 2.5.8}{\geq} \quad \left(\frac{1}{w} \cdot \sum_{j=1}^{j^*_{ii'}-1} w_j \cdot sim_j(t_{ij}, t_{i''j}) \right)$$

$$+ \quad \frac{1}{w} \cdot w_{j^*_{ii'}} \cdot sim_{j^*_{ii'}}(t_{ij^*_{ii'}}, t_{i''j^*_{ii'}})$$

$$+ \quad \left(\frac{1}{w} \cdot \sum_{j=j^*_{ii'}+1}^{j^*_{ii''}-1} w_j \cdot sim_j(t_{ij}, t_{i''j}) \right)$$

$$+ \quad \frac{1}{w} \cdot w_{j^*_{ii''}} \cdot sim_{j^*_{ii''}}(t_{ij^*_{ii''}}, t_{i''j^*_{ii''}})$$

$$+ \quad \left(\frac{1}{w} \cdot \sum_{j=j^*_{ii''}+1}^{N} w_j \cdot sim_j(t_{ij}, t_{i''j}) \right)$$

$$= \quad \frac{1}{w} \cdot \sum_{j=1}^{N} w_j \cdot sim_j(t_{ij}, t_{i''j})$$

$$\overset{Def.\ 2.5.9}{=} \quad Sim(t_i, t_{i''})$$

$$\square$$

Appendix D

Generic Sampling in GenSam

In this appendix, we list the usage message of C program GenSam for documentation reasons. GenSam implements algorithm GENSAM. This program provides all options of generic sampling (see section 4.3.1). The only obligatory arguments to GenSam are the tabfile which contains the focusing input, the domfile which describes information on attribute types and domains, and the outfile which is the name (stem) of the file(s) for the focusing output(s).

The default instantiation of generic sampling in GenSam is simple random sampling with replacement. If users want to apply different sampling techniques, or to prepare the focusing input by sorting or stratification before sampling, they specify the respective options and their values, if they do not use generic sampling in CLEMENTINE. Note, for leader sampling, GenSam supports simultaneous selection of n samples for varying similarity threshold values between d and e.

```
usage: GenSam [-s] [-r#] [-t#] [-m#] [-p] [-z#] [-x#] [-f#]
              [-j#] [-d#] [-e#] [-n#] [-y#] [-w#] [-o]
              [-h hdrfile] tabfile domfile outfile(s)

Generic Sampling

-s          sort table before sampling    (default: no)
-r#         stratification type            (default: 0)
0           no stratification
1           stratification w.r.t. most important attribute
2           stratification w.r.t. all relevant attributes
3           stratification w.r.t. maxsize (option x)
-t#         basic sampling technique       (default: 0)
0           random sampling
1           systematic sampling
2           leader sampling
3           similarity-driven sampling
```

T. Reinartz: Focusing Solutions for Data Mining, LNAI 1623, pp. 281-282, 1999
© Springer-Verlag Berlin Heidelberg 1999

```
-m#        sample size                  (default: automatically)
-p         simulate prototype weighting (default: no weighting)

context-dependent options:
--------------------------
-x#        maximum stratum size         (default: 1)
-z#        discretization method        (default: 0)
 0         equal-width
 1         equal-frequency
 2         fayyad-irani
 3         hong
-f#        start position               (default: random)
-j#        step size                    (default: automatically)
-d#        initial similarity threshold (default: 0.8)
-e#        last similarity threshold    (default: 0.99)
-n#        number of samples            (default: 1)
-y#        adaptation type simsam       (default: 1)
 0         w.r.t. minimum similarity
 1         w.r.t. average similarity
 2         w.r.t. maximum similarity
-w#        relevance weights function   (default: dom file)
 0:        information gain
 1:        information gain ratio
 2:        sym. info. gain ratio
 3:        gini index
 4:        normalized gini index
 5:        relevance
 6:        chi^2-measure
 7:        log2(g)/N
 8:        red. of desc. length 1
 9:        red. of desc. length 2
10:        red. of desc. length 3
11:        specificity 1 gain
12:        specificity 1 gain ratio
13:        sym. spec. 1 gain ratio
14:        specificity 2 gain
15:        specificity 2 gain ratio
16:        specificity 3 gain
17:        specificity 3 gain ratio
18:        <empty>
-o         overwrite weights in dom file (default: no overwrite)

file (options):
----------------
-h          read table header (column names) from hdrfile
hdrfile     file containing table header (column names)
tabfile     table file to read (first line containing column names)
domfile     file containing domain definitions
outfile(s)  file (stem) to write sample(s) to
```

Appendix E

More Experimental Results

In this appendix, we present more experimental results. We show bar charts that contain the same information as bar charts in section 6.2 but with different separations on the x axis and different patterns for bars (i.e., if the original bar chart in section 6.2 separates focusing contexts on the x axis and includes patterns for focusing solutions, the corresponding bar chart here separates focusing solutions on the x axis and includes patterns for focusing contexts, as well as vice versa, respectively). Moreover, we also present bar charts for average and maximum evaluation values rather then minimum values as in section 6.2 for both evaluation criteria E_F and E_W. These additional bar charts enable comparisons between different focusing solutions beyond their optimal results. Table E.1 depicts a complete list of bar charts. For each bar chart, this table indicates the evaluation criterion, whether patterns of bars refer to focusing contexts or focusing solutions (and whether then the x axis separates focusing solutions or focusing contexts, respectively), figure number, and page number.

Table E.1: More Experimental Results

Evaluation	Bars	Figure			Page		
		Min	Avg	Max	Min	Avg	Max
E_F	solutions	6.2	E.2	E.4	183	285	287
E_F	contexts	E.1	E.3	E.5	284	286	288
E_W for C4.5	solutions	6.3	E.7	E.9	189	290	292
E_W for C4.5	contexts	E.6	E.8	E.10	289	291	293
E_W for IB	solutions	6.4	E.12	E.14	196	295	297
E_W for IB	contexts	E.11	E.13	E.15	294	296	298
E_F vs. E_W for C4.5	solutions		E.16			299	
E_F vs. E_W for C4.5	contexts		6.5			202	
E_F vs. E_W for IB	solutions		E.17			300	
E_F vs. E_W for IB	contexts		6.6			209	
E_W for C4.5 vs. E_W for IB	solutions		E.18			301	
E_W for C4.5 vs. E_W for IB	contexts		6.7			216	

T. Reinartz: Focusing Solutions for Data Mining, LNAI 1623, pp. 283-301, 1999
© Springer-Verlag Berlin Heidelberg 1999

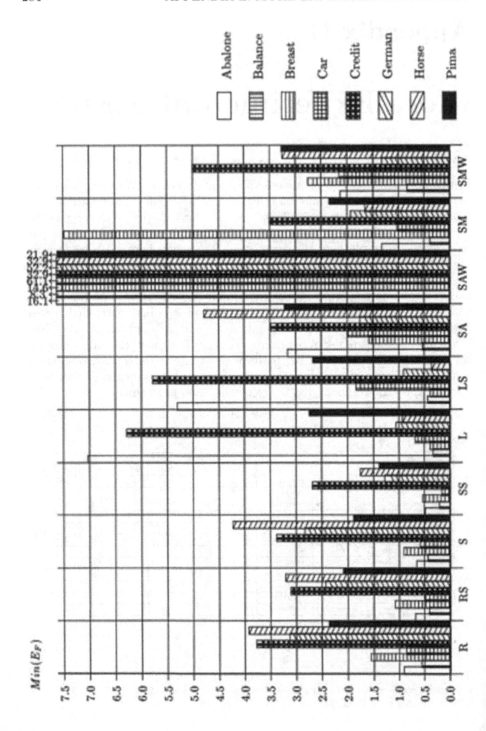

Figure E.1: Minimum Filter Evaluation II

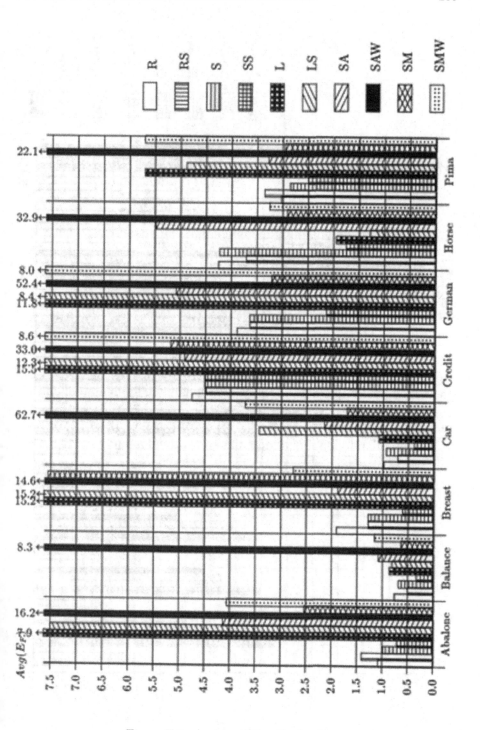

Figure E.2: Average Filter Evaluation

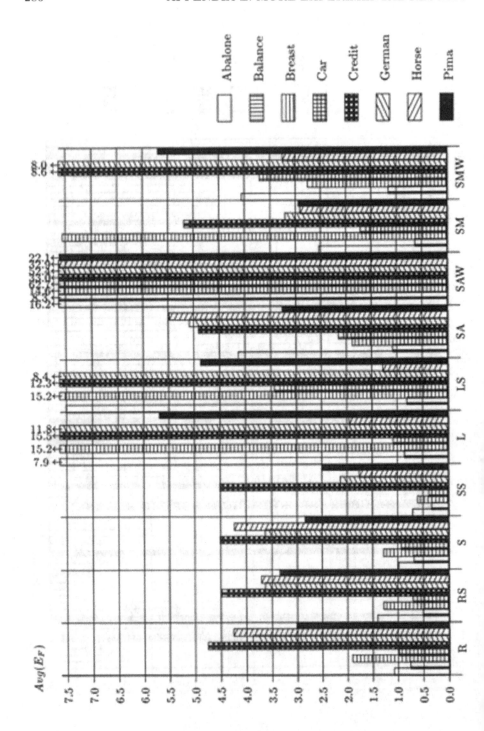

Figure E.3: Average Filter Evaluation II

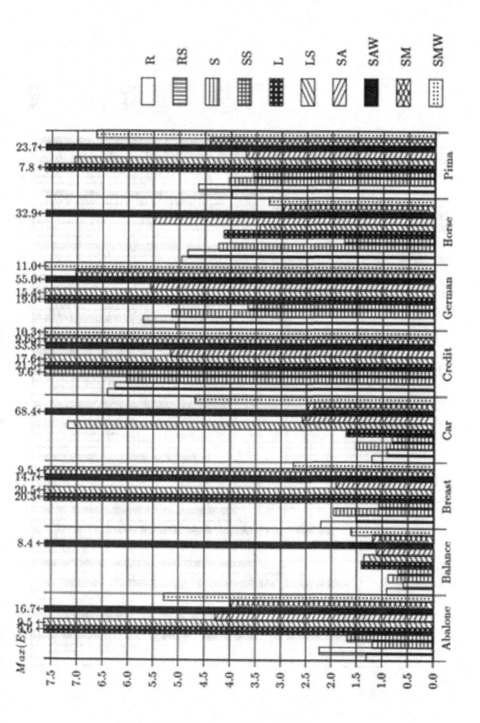

Figure E.4: Maximum Filter Evaluation

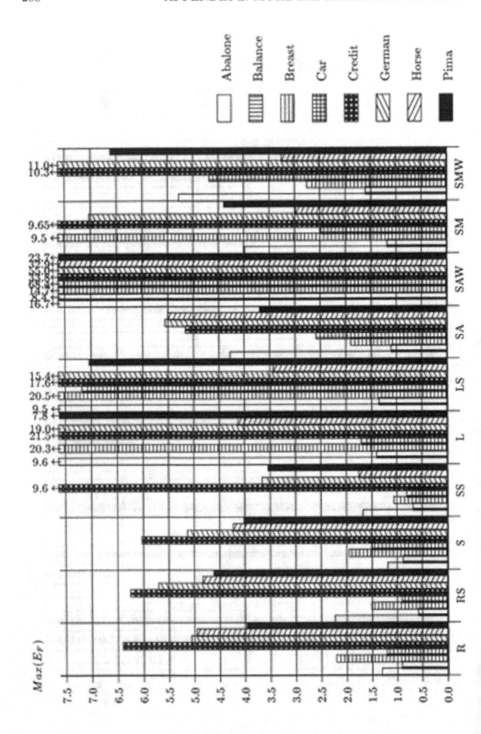

Figure E.5: Maximum Filter Evaluation II

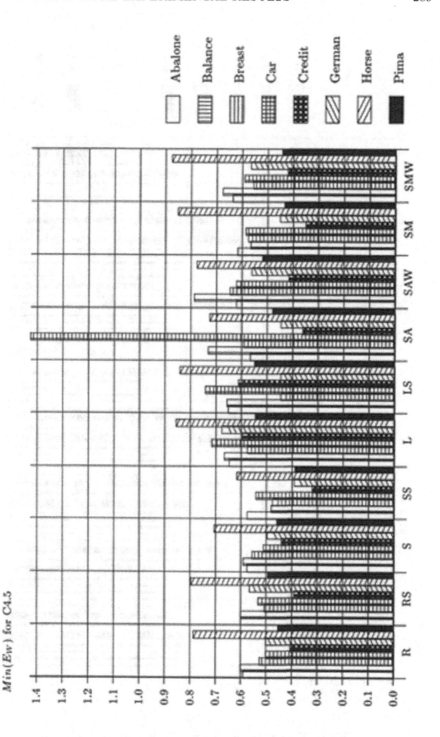

Figure E.6: Minimum Wrapper Evaluation for C4.5 II

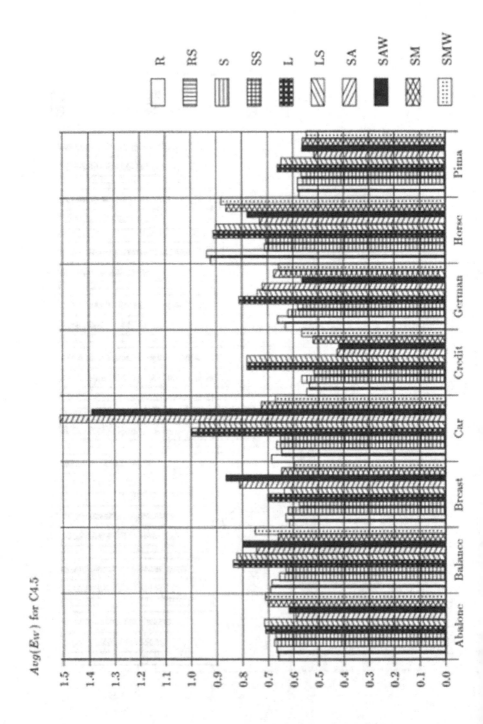

Figure E.7: Average Wrapper Evaluation for C4.5

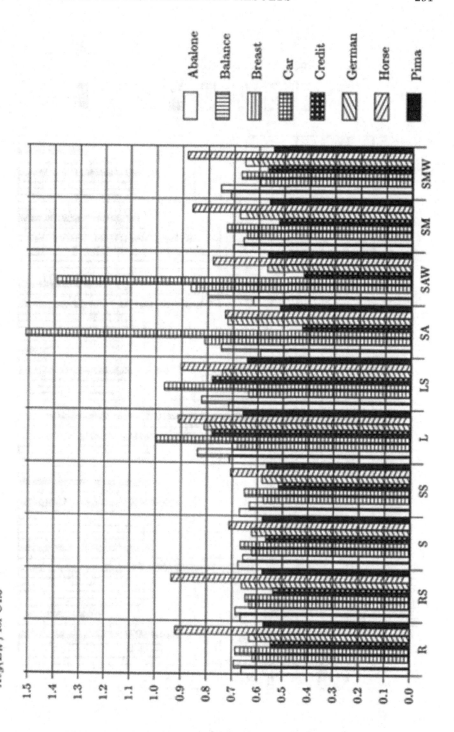

Figure E.8: Average Wrapper Evaluation for C4.5 II

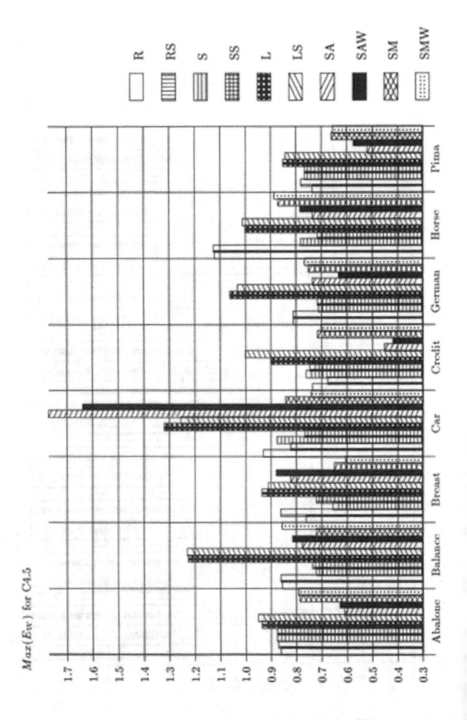

Figure E.9: Maximum Wrapper Evaluation for C4.5

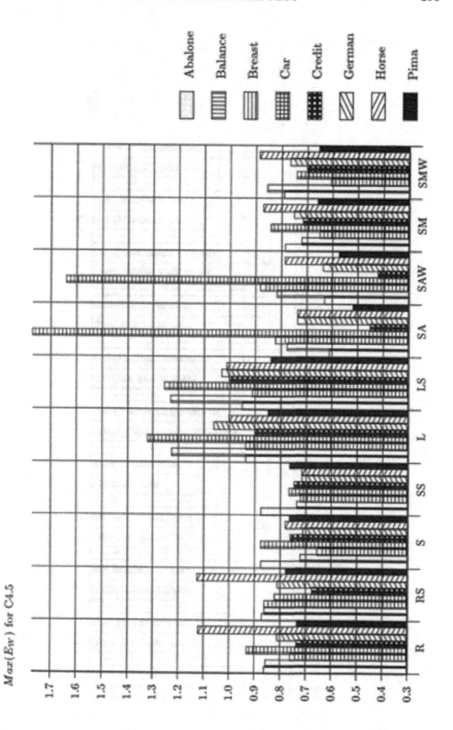

Figure E.10: Maximum Wrapper Evaluation for C4.5 II

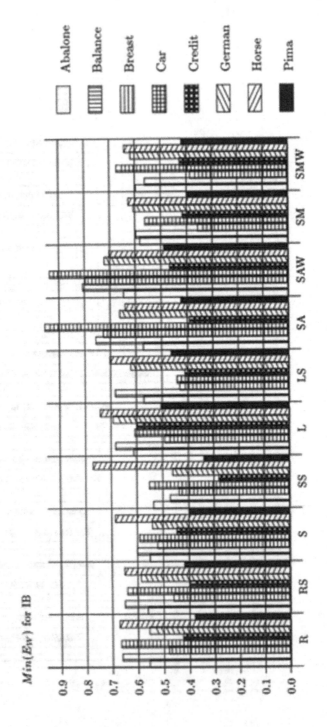

Figure E.11: Minimum Wrapper Evaluation for IB II

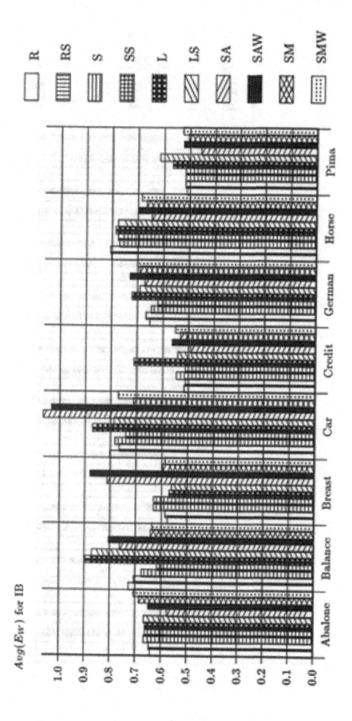

Figure E.12: Average Wrapper Evaluation for IB

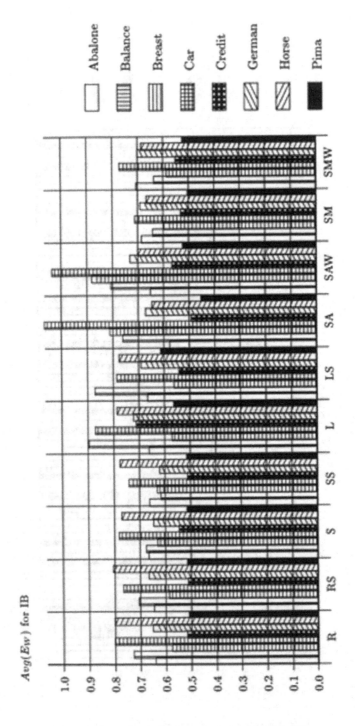

Figure E.13: Average Wrapper Evaluation for IB II

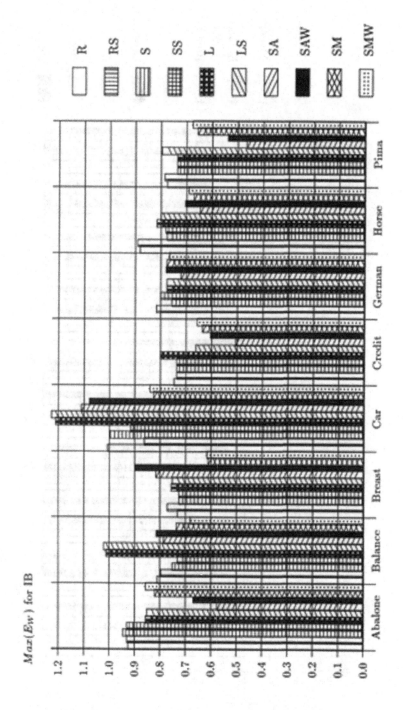

Figure E.14: Maximum Wrapper Evaluation for IB

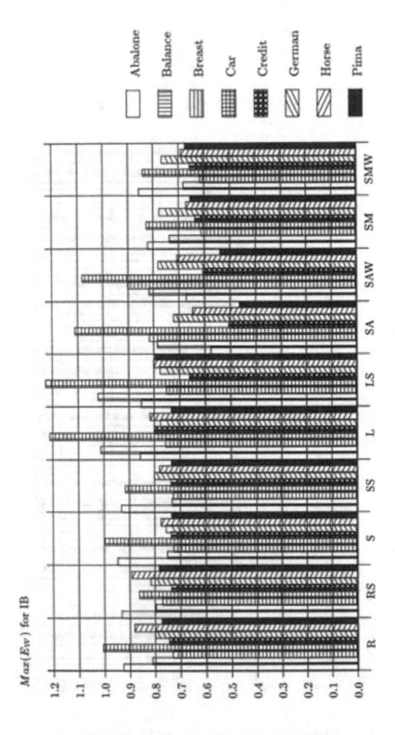

Figure E.15: Maximum Wrapper Evaluation for IB II

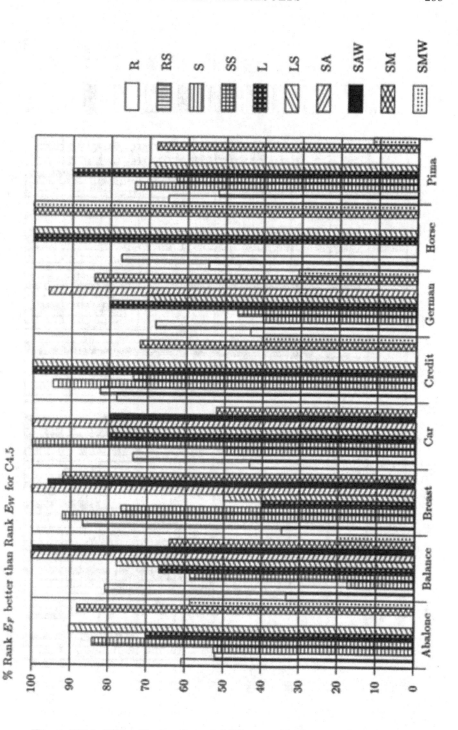

Figure E.16: Filter Evaluation and Wrapper Evaluation for C4.5 II

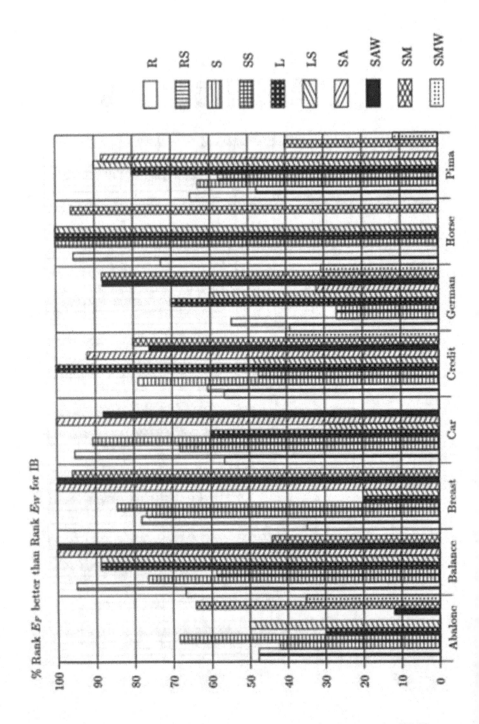

Figure E.17: Filter Evaluation and Wrapper Evaluation for IB II

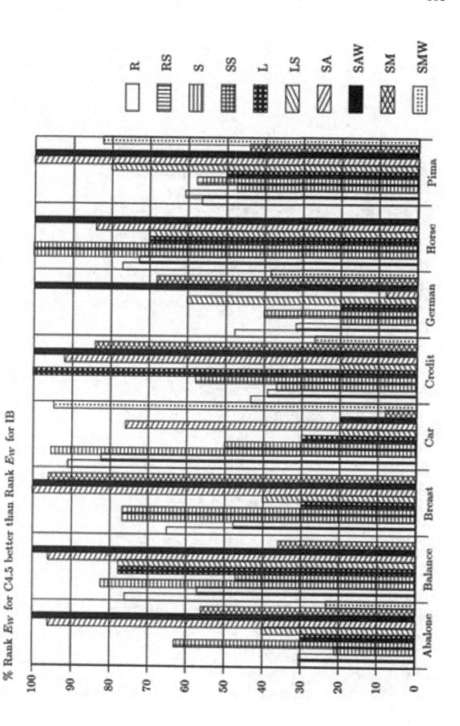

Figure E.18: Wrapper Evaluation for C4.5 and Wrapper Evaluation for IB II

Index

T. Reinartz: Focusing Solutions for Data Mining, LNAI 1623, pp. 303-307, 1999
© Springer-Verlag Berlin Heidelberg 1999

Curriculum Vitae

Personal Details

Name	Thomas Peter Reinartz
Address	Rührweg 2/1
	89081 Ulm-Lehr
Date of Birth	April, 23rd, 1969
Place of Birth	Rheydt
Nationality	German
Marital Status	Married, 2 Children

School Education

Aug. 1975 – June 1979		Gemeinschaftsgrundschule Geistenbeck
		Mönchengladbach
Aug. 1979 – May 1988		Städtisches Gymnasium Odenkirchen
		Mönchengladbach
	May 1988	Allgemeine Hochschulreife

Study

Oct. 1988 – Aug. 1994		Computer Science & Mathematics
		Kaiserslautern University
	Aug. 1994	Diplom

Professions

Oct. 1990 – March 1994		Research Assistant
		German Research Center for AI
		Kaiserslautern
Oct. 1993 – June 1994		Research Assistant
		AG Expert Systems
		Kaiserslautern University
Oct. 1994 – Sept. 1997		Ph.D. Student
		Daimler-Benz AG, Research & Technology
		Ulm
April 1996 – Sept. 1996		Visiting Scholar
		Stanford University
since Oct. 1997		Research Scientist
		Daimler-Benz AG, Research & Technology
		Ulm

T. Reinartz: Focusing Solutions for Data Mining, LNAI 1623, p. 309-309, 1999
© Springer-Verlag Berlin Heidelberg 1999

Lecture Notes in Artificial Intelligence (LNAI)

Lecture Notes in Computer Science